Community Genetics

KARGER

Printed in Switzerland
on acid-free paper by
Reinhardt Druck, Basel

Appears quarterly:
1 volume per year
(4 issues)

Community Genetics

Guidelines for Authors

Submission

Only original papers written in English are considered and should be sent by first class mail or courier to:

Prof. L.P. ten Kate
S. Karger AG
Editorial Office Community Genetics
PO Box
CH–4009 Basel (Switzerland)
Tel. +41 61 306 1356
Fax +41 61 306 1434
E-Mail e.lawrence@karger.ch
www.karger.ch/cmg

Mailing address for courier deliveries only:
Allschwilerstrasse 10
CH–4055 Basel (Switzerland)

Information for Contributors

Community Genetics is multidisciplinary, covering contributions on community genetics from diverse fields such as medical genetics, epidemiology, primary care, community medicine and public health education, and also includes psychological, social, legal and ethical aspects. It features original research papers, reviews, short communications, case, country and other reports, as well as journal club, news, opinion and correspondence sections.

Manuscripts should be submitted in quadruplicate (with four sets of illustrations of which three are originals), typewritten double-spaced on one side of the paper, with a wide margin and pages numbered. *Lines should also be numbered in the margin.* The final version of your manuscript and illustrations should also be submitted on disk; final high-resolution pictures need only be sent after acceptance of the paper together with a soft copy on disc. For details refer to 'Disk Submission' or www.karger.com/journals/disksub.htm. Editorial review and copy-editing will be executed from the disk version of your manuscript. Names and addresses of four experts in the appropriate area of research should accompany each manuscript.

Conditions

All manuscripts are subject to editorial review. All manuscripts must be accompanied by a covering letter indicating the section for which the paper is intended and signed by the first and last authors who should give assurance that the data contained in the manuscript are not under simultaneous consideration by any other journal. Submission of an article for publication implies the transfer of the copyright from the author to the publisher upon acceptance. Accepted papers become the permanent property of *Community Genetics* and may not be reproduced by any means, in whole or in part, without the written consent of the publisher. It is the author's responsibility to obtain permission to reproduce illustrations, tables, etc. from other publications.

General Arrangement

Title Page: The first page of each paper should indicate the title, the authors' names, the institute where the work was conducted, and a short title for use as running head. NB: Authors wishing to preserve the phonetic meaning of diacritics (PubMed reduces diacritics to their root characters) must spell their names accordingly when submitting manuscripts (e.g. Müller should be Mueller).

Full Address: The exact postal address of the corresponding author complete with postal code must be given at the bottom of the title page. Please also supply phone and fax numbers, as well as E-Mail address.

Key Words: For indexing purposes, a list of 3–10 key words in English is essential.

Abstract: Each paper needs an abstract of up to 10 lines.

Footnotes: Avoid footnotes. When essential, they are numbered consecutively and typed at the foot of the appropriate page.

Tables and illustrations: Tables and illustrations (both numbered in Arabic numerals) should be prepared on separate sheets. Tables require a heading and figures a legend, also prepared on separate pages. For the reproduction of illustrations, only good drawings and original photographs can be accepted; negatives or photocopies cannot be used. Due to technical reasons, figures with a screen background should not be submitted. When possible, group several illustrations on one block for reproduction (max. size 180 × 223 mm) or provide crop marks. On the back of each illustration, indicate its number, the author's name, and 'top' with a soft pencil. Electronically submitted b/w half-tone and color illustrations must have a final resolution of 300 dpi after scaling, line drawings one of 800–1,200 dpi.

Color Illustrations: Up to 6 color illustrations per page can be integrated within the text at the special price of CHF 660.–/USD 545.00 per page. Color illustrations are reproduced at the author's expense.

References: In the text, identify references by Arabic numerals [in square brackets]. Material submitted for publication but not yet accepted should be noted as 'unpublished data' and not be included in the reference list. The list of references should include only those publications which are cited in the text. Do not alphabetize; number references in the order in which they are first mentioned in the text. The surnames of the authors followed by initials should be given. There should be no punctuation other than a comma to separate the authors. Preferably, please cite all authors. Abbreviate journal names according to the *Index Medicus* system. (Also see International Committee of Medical Journal Editors: Uniform requirements for manuscripts submitted to biomedical journals. N Engl J Med 1997;336:309–315. www.icmje.org)

Examples

(a) Papers published in periodicals: Chatel J-M, Bernard H, Orson FM: Isolation and characterization of two complete Ara h 2 isoforms cDNA. Int Arch Allergy Immunol 2003;131:14--18.

(b) Papers published only with DOI numbers: Theoharides TC, Boucher W, Spear K: Serum interleukin-6 reflects disease severity and osteoporosis in mastocytosis patients. Int Arch Allergy Immunol DOI: 10.1159/000063858.

(c) Monographs: Matthews DE, Farewell VT: Using and Understanding Medical Statistics, ed 3, revised. Basel, Karger, 1996.

(d) Edited books: DuBois RN: Cyclooxygenase-2 and colorectal cancer; in Dannenberg AJ, DuBois RN (eds): COX-2. Prog Exp Tum Res. Basel, Karger, 2003, vol 37, pp 124–137.

Digital Object Identifier (DOI)

S. Karger Publishers supports DOIs as unique identifiers for articles. A DOI number will be printed on the title page of each article. DOIs can be useful in the future for identifying and citing articles published online without volume or issue information. More information can be found at www.doi.org.

Research Papers

Full papers as well as short communications are arranged as follows: Introduction, Methods, Results, Discussion. Their abstracts are also structured with subheadings as follows: Background/Aims/Objective(s), Methods, Results, Conclusion(s).

Case Reports

Case reports describe illustrative projects in the field of community genetics, e.g design and progress of an educational program or the protocol and achievement of a genebank. Case reports describing individual patients are not accepted.

Country Reports

Country reports review genetic diseases, (genetic) communities and community genetics in a particular country, or part of a country. Papers are preferably multi-authored. The primary author is encouraged to obtain input from the country's or region's health department and from patient or other advocacy organizations. Papers should follow the same format:

1. An *introduction* with data on the country's geography, population size, political system, official language(s) and health care system.
2. A section on the *population history* of the country: founders, migrants, bottlenecks, etc.
3. A section on present *population substructure* (stratifications), including a description of (genetic) communities and their reproductive characteristics (e.g. consanguinity).
4. A review of *genetic diseases* in the country, with reference to population characteristics.
5. A review of what is available to *support* patients, families and communities with genetic diseases, including genetic diagnosis and counseling, genetic screening, genetic interest groups, governmental involvement, and public and professional education.
6. A review of *unmet needs*, including existing barriers and how they could be overcome.

The text should not exceed 3,000 words, the number of tables and/or figures should not exceed five, and the number of references should be 50 or less. Suggested title: Country Name: Communities and Community Genetics. The abstract should not exceed 250 words.

Proofs

Unless indicated otherwise, proofs are sent to the first-named author and should be returned with the least possible delay. Alterations made in proofs, other than the correction of printer's errors, are charged to the author. No page proofs are supplied.

Reprints

Order forms and a price list are sent together with the proofs. Orders submitted after the issue is printed are subject to considerably higher prices.

KARGER

Fax +41 61 306 12 34
E-Mail karger@karger.ch
www.karger.com

The Guidelines for Authors are available at:
www.karger.com/cmg_Guidelines

Community Genetics

General Information

Publication data: 'Community Genetics' is published 4 times a year. Volume 8 with 4 issues appears in 2005.

Subscription rates: Subscriptions run for a full calendar year. Prices are given per year.

Personal subscription:

Print or Online	Print/Online combined
CHF 96.00	CHF 137.00
EUR 71.00	EUR 101.00
USD 78.00	USD 111.00
postage and handling	(added to print and print/online)
CHF 21.80 Europe,	CHF 32.40 Overseas
EUR 15.80	
USD 21.80	

Institutional subscription:

Print or Online	Print/Online combined
CHF 756.00	CHF 816.00
EUR 560.00	EUR 604.00
USD 615.00	USD 663.00
postage and handling	(added to print and print/online)
CHF 27.20 Europe,	CHF 40.40 Overseas
EUR 19.60	
USD 27.20	

Airmail surcharge: CHF 27.20 / USD 22.60

Microforms: Please direct all enquiries and orders to:
University Microfilms Inc.
300 North Zeeb Road, P.O. Box 1346
Ann Arbor, MI 48106-1346 (USA)

Subscription orders: Orders can be placed at agencies, bookstores, directly with the Publisher
S. Karger AG, P.O. Box
CH–4009 Basel (Switzerland)
or with any of the following distributors:
Deutschland: S. Karger GmbH, Postfach
D–79095 Freiburg
France: Librairie Luginbühl
36, bd de Latour-Maubourg
F–75007 Paris
Great Britain, Ireland: S. Karger AG
4 Rickett Street
London SW6 1RU, UK
USA: S. Karger Publishers, Inc.
26 West Avon Road, P.O. Box 529
Farmington, CT 06085, USA
India, Bangladesh, Sri Lanka:
Panther Publishers Private Ltd.
33, First Main
Koramangala First Block
Bangalore 560 034, India
Japan: Karger Japan, Inc.
Yushima S Bld. 3F
4-2-3, Yushima, Bunkyo-ku
Tokyo 113-0034, Japan

Change of address: Both old and new address should be sent to the subscription source.

Single issues and back volumes: Information can be obtained through the Publisher.

Advertising: Correspondence and rate requests should be addressed to the Publisher.

Bibliographic indices: This journal is regularly listed in bibliographic services, including *Current Contents®* and Index Medicus/MEDLINE.

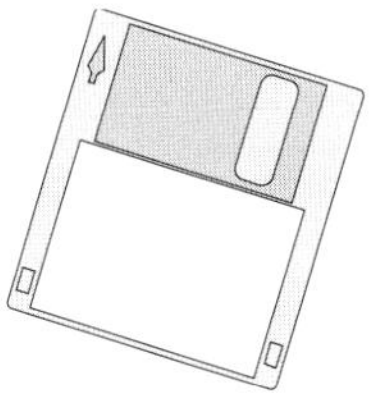

Disk Submission

A **comprehensive version** of these guidelines for the **submission of manuscripts and illustrations on disk** can be accessed online under **www.karger.com/disksubmission** or requested as a leaflet from the editorial department.

Please follow the general rules on style and arrangement in the **'Guidelines for Authors'** of the journal. Submit a **double-spaced printout** of your manuscript matching the file on disk exactly.

Technical Information
Store only data pertinent to this manuscript. Include your fonts (screen and printer fonts for Macintosh). For a **text only** manuscript, use a 3.5-inch disk. For **text and illustrations,** use 3.5 or 5.25 Magneto Optical Disc; CD-ROM (DOS or hybrid); iomega Zip/DOS (100 MB), or 44 MB, 88 MB or 200 MB SyQuest. Compress extensive data using ZIP format.

Data Description
Label your disk with the **title of the journal or book** for which it is submitted; a **short version** of the paper's title; your **name,** postal and e-mail address, fax and phone number; **hard- and software** used, including the version; and the **complete list of the file names.** File names consist of eight characters before the dot identifying the file and three characters after the dot identifying the application.

Text and Format
The **preferred word-processing package is MS-Word.** Other commonly used PC text programs and ASCII formats are accepted.
Update your data when you revise subsequent to the reviewer's or editor's decision and mark revisions clearly. Manuscripts accepted for publication are subject to **copy editing:** changes and/or questions will be marked in the manuscript, which will be returned to you with your disk and the galley proofs for checking.
Enter your text continuously flush left. Do not use any layout functions. Do not split words at the end of a line. **Do not use indentations.** Use a blank line before and a hard return at the end of paragraphs. **Arrange headings flush left.** Insert only **one space after punctuation marks.** Insert Greek letters, mathematical symbols etc. with your word-processing program. Use **boldface and italics** as well as sub- and superscript where appropriate.
Tables are part of the text and should be placed at the end of the text file. Separate the individual parts of a table by lines. Use the tabulator instead of spaces. Format operation data on one line in the same column.

Legends to illustrations are part of the text and should also be listed at the end of the text file.

Illustrations
Store illustrations as separate files: do not integrate them in your text. Export **line drawings/vector graphics** as TIF, EPS or WMF format. Use Photoshop for processing and retouching **scanned half-tone images.** Save the original scan and the processed version. Export **black and white or color images** as TIF or EPS format in their anticipated size in print. **For detailed information** on how to submit illustration data on disk please consult the leaflet or **www.karger.com/disksubmission.**

KI 02 DISK

KARGER

Fax +41 61 306 12 34
E-Mail karger@karger.ch
www.karger.com

The Journal Home Page is available at:
www.karger.com/cmg

Journal Publication

Medical Statistics

Current Developments in Statistical Methodology for Genetic Architecture of Complex Diseases

Editors
Jürg Ott
Derek Gordon

Genetics; Epidemiology, Human Genetics, Medical Statistics

This publication provides an overview of current advances in statistical methodology for gene mapping and parameter estimation in complex genetic traits. Presenting selected papers from the conference 'Current Developments in Statistical Methodology for Genetic Architecture of Complex Diseases' held in Oberwolfach in February 2003, it discusses topics such as statistical population genetics, novel approaches for positional gene identification, statistical modelling of multiple disease loci, haplotype estimation and haplotype association analysis, novel designs for nuclear family studies, whole genome analysis, and analysis of large pedigrees.
A valuable single source of information, this collection of reviews will interest statisticians, geneticists, epidemiologists, physicians studying complex traits, and anyone in the scientific community interested in statistical methodology applied to questions that arise in mapping genes for complex traits.

Contents

www.karger.com/hhe

KI 04214

Human Heredity
Editor: Ott, J. (New York, N.Y.)
ISSN 0001–5652
Special issue: Vol. 56, No. 1–3 (2003)
Included in subscription

Medical Statistics
Current Developments in Statistical Methodology for Genetic Architecture of Complex Diseases
Meeting at the 'Mathematisches Forschungsinstitut Oberwolfach', Germany, February 2003
Editors: Ott, J.; Gordon, D. (New York, N.Y.)
IV + 150 p., 32 fig., 41 tab., soft cover, 2003
CHF 45.– / EUR 32.– / USD 39.25
Prices subject to change
EUR price for Germany, USD price for USA only
ISBN 3–8055–7676–5

KARGER

Order Form

Please send: ____ copy/ies
Postage and handling free with prepayment

Payment:
☐ Check enclosed ☐ Please bill me

Please charge to my credit card
☐ American Express ☐ Diners ☐ Eurocard
☐ MasterCard ☐ Visa

Card No.: ____________

Exp. date: ____________

Orders may be placed with any bookshop, subscription agency, directly with the publisher or through a Karger distributor.

Fax: +41 61 306 12 34
S. Karger AG, P.O. Box, CH–4009 Basel (Switzerland)
E-Mail orders@karger.ch, **www.karger.com**

Name/Address: ____________

Date: ____________

Signature: ____________

Community Genetics

Community Genetics in the Arabian Gulf Region

Guest Editor

Lihadh Al-Gazali, Al Ain

6 figures and 22 tables, 2005

KARGER

Basel · Freiburg · Paris · London · New York · Bangalore · Bangkok · Singapore · Tokyo · Sydney

S. Karger
Medical and Scientific Publishers
Basel • Freiburg • Paris • London
New York • Bangalore • Bangkok
Singapore • Tokyo • Sydney

Drug Dosage
The authors and the publisher have exerted every effort to ensure that drug selection and dosage set forth in this text are in accord with current recommendations and practice at the time of publication. However, in view of ongoing research, changes in government regulations, and the constant flow of information relating to drug therapy and drug reactions, the reader is urged to check the package insert for each drug for any change in indications and dosage and for added warnings and precautions. This is particularly important when the recommended agent is a new and/or infrequently employed drug.

Printed in Switzerland on acid-free paper by
Reinhardt Druck, Basel
ISBN 3–8055–7855–5

KARGER

Fax +41 61 306 12 34
E-Mail karger@karger.ch
www.karger.com

Vol. 8, No. 1, 2005

Community Genetics

Contents

KARGER

Fax +41 61 306 12 34
E-Mail karger@karger.ch
www.karger.com

Access to full text and tables of contents, including tentative ones for forthcoming issues: www.karger.com/cmg_issues

Editorial

Community Genetics

Community Genet 2005;8:5–6
DOI: 10.1159/000083329

Genetics: Global Challenge, Regional Focus

Advances in Community and Preventive Genetics

This issue of *Community Genetics* is devoted to the scientific content of the International Genetics Congress that was held in December 2003 in Dubai, United Arab Emirates. The aim of the Congress was to bring together the global genetic community to hear, share and discuss a broad range of genetic-related issues, with the aim of setting improved standards in education, service and research.

The Congress entitled *Genetics-Global Challenge, Regional Focus* was organized by the Genetics and Development Research Priority Group at the Faculty of Medicine, United Arab Emirates University, Al Ain. Special emphasis was given to advances in community and preventive genetics, and most scientific contributions thus focused on various aspects of genetic disorders from a global perspective, and on the link between basic scientific research and clinical genetics as a bridge to community health.

In compiling this issue, the aim of the invited editors was to focus on common genetic problems in Arab countries and the Middle East, and on the development of preventive strategies in this region. Consanguineous marriage is common in the Arab world, reaching rates as high as 40–50%. This has wide-ranging implications not only for increased rates of recessive disorders and lethal malformations [Dawodu et al.], but also has an impact on other disease conditions, such as the risk of cancer in individuals and populations. Certain specific genetic conditions prevalent in the Arabian Gulf region are presented and defined [Baysal; Rajab et al.; Teebi and Teebi]. There has already been substantial integration of information and databases for Arab populations [Teebi and Teebi], but this progress needs to be further considered in the context of genetic diversity among different Arab groups. Therefore, further characterization of the prevailing genetic disorders in each geographic location, together with their mutations, is needed to plan for appropriate screening and testing protocols [Samilchuk]. However, the presence of multiple discrete subpopulations creates practical difficulties in the organization of national or regional screening programs for genetic disorders in this region [Bittles]. With rapidly changing demographic profiles and the increasing global impact of inherited diseases, a better understanding of the influence of endogamy and consanguinity on population structure and transmission of diseases is also crucial in planning educational counseling and screening programs [Bittles].

Along the same line, the meeting addressed issues related to the challenges associated with preventive genetics in Islamic countries, and novel strategies adopted to overcome these problems, such as premarital screening and pre-implantation diagnostics [Ozand et al.; Al Arrayed]. In this regard, the specific attitudes and perceptions of families, societies and cultural groups towards genetic disorders and genetic counseling need to be taken into account [Al-Gazali].

Mechanisms underlying gene-environment interactions are considered in the context of developing appropriate preventive and intervention strategies [Eapen et al.]. Hassan et al. present a new model to study the genetics of complex diseases using multigenerational highly inbred Arab pedigrees. These extensive Arab pedigrees

KARGER

Fax +41 61 306 12 34
E-Mail karger@karger.ch
www.karger.com

1422–2795/05/0081–0005$22.00/0

Accessible online at:
www.karger.com/cmg

will provide a powerful resource for the study of common genetic disorders like hypertension, diabetes and obesity.

In this special issue, Prof. ten Kate argues that *Community Genetics* is the bridge between Clinical Genetics and Public Health, while Prof. Raeburn suggests possible funding models for the management of genetic conditions on a worldwide basis, with the aim of achieving global health care.

The invited editors thank Prof. ten Kate, the Editor of *Community Genetics*, and S. Karger AG, the publishers, for their strong and enthusiastic support in the production of this special issue of *Community Genetics* devoted to the proceedings of the Dubai International Genetics Congress, and Prof. Allan Bittles for supervising the procedures of peer review. The organizers of the meeting would like to acknowledge the encouragement and support of the United Arab Emirates University and particularly H.E. Sheikh Nahayan Mabarak Al-Nahayan, Minister of Higher Education and Scientific Research and Chancellor. The conference was also supported by the Sheikh Hamdan Bin Rashid Al Maktoum Award for Medical Sciences, the Center for Arab Genomic Studies, the Emirates Medical Association Pediatric Society, Abbott Diagnostics, Affymetrix, Al-Khayat Scientific Supplies, Al-Nawras, Nanogen, Beckman Coulter, HVD-Life Sciences, Philips, Metabion, Nestle and Hoffmann-La-Roche.

Further information on the meeting is available on the following website: http://www.alain-genetics.com.

Lihadh Al-Gazali, Guest Editor
Valsa Eapen, Associate Guest Editor

Genetics and Development Research Priority Group
Faculty of Medicine and Health Sciences
United Arab Emirates University

Community Genetics

Community Genet 2005;8:7–11
DOI: 10.1159/000083330

Community Genetics: A Bridge between Clinical Genetics and Public Health

Leo P. ten Kate

Department of Clinical Genetics and Human Genetics, VU University Medical Center, Amsterdam, The Netherlands

Key Words
Clinical genetics · Counseling · Genetic communities · Public health

Abstract
Clinical genetics and public health differ in their subjects; in who takes the initiative; in their aims, methods and outcome measures, and therefore in their ability to handle sensitive issues. They are seemingly incompatible. Nevertheless, they can be reconciled in community genetics. Community genetics combines the best from clinical genetics and public health. It shares with clinical genetics its aim of empowerment, its methods of information and support, its outcome measures and its possibility to handle sensitive issues. It has in common with public health its subjects; the origin of its initiatives, and non-reproductive prevention. In general, communities can be defined at different levels: geographically, linguistically, socially or genetically. Genetic communities can also be defined at different levels: on the basis of common heritage, on the basis of common practice, or on the basis of common problems.

Introduction

A bridge is a structure carrying a road, path, railway, etc. across a river, ravine, road, railway or another obstacle. Here I refer to a second meaning of the word 'bridge': something which is intended to reconcile or form a connection between two seemingly incompatible things [1]. So I pretend to show here that clinical genetics and public health are two seemingly incompatible things and that community genetics can reconcile them, as illustrated in figure 1.

Clinical Genetics and Public Health Are Two Worlds Apart

Clinical genetics and public health differ in many respects. They differ in their subjects; in who takes the initiative; in their aims, methods and outcome measures, and therefore in their ability to handle sensitive issues.

Clinical genetics is concerned with individual persons or couples or families who have or fear a health problem. These individuals, couples or families are seen one by one. Public health, on the other hand, is not primarily interested in persons or families who are already aware of a problem, but it focuses on people who may not yet be aware but are at risk of developing a health problem. Where possible, a group approach is preferred above an individual one.

KARGER

Fax + 41 61 306 12 34
E-Mail karger@karger.ch
www.karger.com

1422–2795/05/0081–0007$22.00/0

Accessible online at:
www.karger.com/cmg

Prof. Leo P. ten Kate, MD, PhD
Department of Clinical Genetics and Human Genetics
VU University Medical Center, POB 7057
NL–1007 MB Amsterdam (The Netherlands)
Tel. +31 20 4440786, Fax +31 20 4444181, E-Mail lp.tenkate@vumc.nl

In clinical genetics, the initiative comes from the patient or the family. The health care provider sits in his office waiting for the worried patients and clients. In public health, the initiative comes from the provider, who, following the analogy, leaves his office in search of people who might need his help and advice.

The aim of public health is to promote health and prevent disease. Although these are laudable intentions, clinical genetics has another goal, namely empowerment of the counselee, so that they can understand their situations, risks and options and make there own choices according to their own preferences and values (see Appendix) [2]. This leads to a non-directive approach, which is especially important for sensitive issues and difficult decisions like refraining from having offspring, adoption, use of gamete donation, or selection of offspring through prenatal diagnosis or pre-implantation genetic diagnosis.

In order to promote health and prevent disease, it may not be enough to inform and support, but a bit of pressure or even coercion may be needed. Public health tries to convince by persuasion and in some situations even mandates certain measures. Clinical genetics also may try to convince in situations were people at risk can obtain greater health benefits, e.g. with frequent surveillance in the case of hereditary cancers. But historically the methodology of clinical genetics has been restricted to informing and supporting the counselee.

In line with the different aims of clinical genetics and public health, these disciplines also differ in their outcome measures. Empowerment can be measured by measuring improvement in perceived personal control [3] or informed choice [4]. Achievements in public health are measured with uptake and compliance, with decrease in numbers affected and also with the amount of economic benefit.

Clinical genetics has the ability to handle sensitive issues, such as reproductive choices and pre-symptomatic testing at the request of the counselee for adult diseases for which there is no cure. If public health enters this field the distinction with eugenics becomes blurred. In these situations, persons who are tested and persons who do not want to be tested run the risk of becoming stigmatized or discriminated against, especially if the testing is done for public health's sake. The same applies when reproductive matters, in the name of prevention, are brought into the sphere of public health. Khoury et al. [5] wrote: 'Assuring the availability and voluntary access to prenatal services and evaluating the impact of these services would fall within the purview of public health practice; nevertheless, although the public health community should take an active role in promoting the use of genetic tests and services when there are proven and cost-effective interventions to prevent disease, death or disability (e.g. phenylketonuria or PKU), public health agencies cannot play a role in promoting the use of genotypic prevention as a method to improve the public's health.' Although we may add that evaluating the impact of prenatal services should be according to the standards of clinical genetics and not according to the standards of public health, the message is clear: reproductive issues are outside the field of public health and should remain there.

The qualification 'genotypic' prevention, comes from Juengst [6], who argues that one should distinguish between genotypic prevention (or geno-prevention) and phenotypic prevention (or pheno-prevention). Genotypic prevention describes efforts to avoid the transmission of particular genotypes to the next generation through reproductive counseling, carrier testing, prenatal diagnosis, and pregnancy termination, while phenotypic prevention describes medical efforts to forestall the clinical manifestation of a genetic disease in an at-risk person. The intervention to prevent PKU in the above statement of Khoury et al. [5] is a clear example of phenotypic prevention. Throughout his paper Juengst [6] makes it clear that genotypic prevention should stay in the domain of the individual couple and should not become a goal of health care providers or a kind of public health policy.

Notwithstanding the good intention to clarify the distinctions, I disagree on the introduction of the term genotypic prevention, since the word 'prevention' suggests that this genotypic prevention is a good and laudable thing. I would prefer not to use the word prevention at all when interruption of genetic trait transmission through reproductive selection is at stake. We should restrict the word prevention to situations where we try to improve the conditions for living creatures, and not use the term for interventions that end the existence of these individuals. Although Alzheimer's disease is a disease of old age, no one would suggest that the best prevention of Alzheimer's disease would be to prevent people to reach old age. Such a suggestion would meet abhorror and should at least be regarded as a perversion of the concept of prevention. By analogy we should abandon the term prevention when it comes to prenatal selection and use the term 'reproductive choice' instead.

Although the word prevention is used frequently outside medicine and more loosely there, within medicine we should stick to clear definitions. Primary prevention means that disease is prevented at all; secondary prevention implies that primary prevention was not possible

Table 1. Examples of measures of primary and secondary prevention and reproductive choice concerning genetic diseases

	Prevention	Reproductive choice
Primary	Radiation protection	Refraining from children
Secondary	Neonatal screening for PKU Surveillance of at-risk persons in families with hereditary cancers	Prenatal diagnosis and termination of pregnancy

or successful, but that the disease can be found at an early, or asymptomatic stage and that there is some intervention available to block or delay further development of symptoms. By analogy we can distinguish primary reproductive choice and secondary reproductive choice [7]. Primary reproductive choice means that there will not be offspring at all: secondary reproductive choice relates to the selection of embryos and fetuses. Examples of all four possibilities that arise by combining primary and secondary with prevention and reproductive choice are shown in table 1.

In conclusion, it is clear that clinical genetics and public health are two seemingly incompatible things. Now I have to show that community genetics can reconcile them.

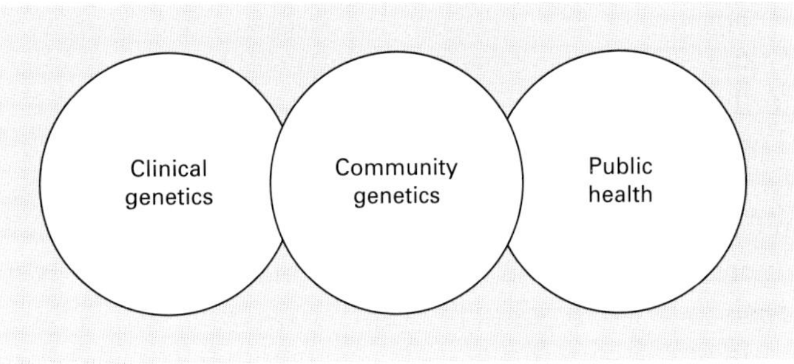

Fig. 1. Community genetics reconciles clinical genetics and public health.

Community Genetics Can Reconcile Clinical Genetics and Public Health

Community genetics can be defined as the science and practice of maximizing the benefits of medical genetics for as many people as possible, while minimizing harm and discomfort from genetics for all. Community genetics combines the best from clinical genetics and public health. It shares with clinical genetics its aim of empowerment, its methods of information and support, its outcome measures and its ability to handle sensitive issues. On the other hand it has in common with public health its subjects, the origin of its initiatives, and non-reproductive prevention. Its emphasis is on autonomy, informed choice, empowerment, beneficence, non-maleficence, and justice, but NOT on prevention as such, health promotion, financial profit, gene pool or future generations.

Genetic screening offers many opportunities to illustrate the difference between an approach directed at diminishing the number of affected persons and an approach aiming to increase informed reproductive choice. As an example I take the case of screening for carriers of cystic fibrosis [8–12]. Those who strive at lowering the birth prevalence of the disease will choose a turnstile opportunity where all members of the target population pass through. Neonates and pregnant women are in such a position where they can be captured easily. Of these two possibilities pregnancy is the opportunity of choice since there is an immediate interest for the women, resulting in a high uptake of testing and a high compliance to proceed with the follow-up procedure. Obtaining formal informed consent should and can be done quickly.

Those who offer carrier screening to increase reproductive choice will choose a moment which allows ample time for decision making and which guarantees the maximum number of reproductive options. In order to be sure that people who do not want to be tested can escape, they will ask for an initiative of the members of the target population, such as coming to a separate appointment for pretest counseling. They are not in a hurry to obtain informed consent. Pre-conceptional couple carrier screening can fulfill these conditions.

The first mention of community genetics in the medical literature was in a paper by Bernadette Modell in 1990 [13]. However, even prior to that, many individuals had already devoted a large part of their lives to applying genetic approaches at the community level [14]. It was, until 1990, a concept without a name. Once given a name, communication about the concept became much easier. The first issue of the international scientific journal *Community Genetics* appeared in 1998 and from 1999 on-

Table 2. Examples of differently defined communities

Criterion	Examples
Geography	Village, quarter, city, region, country
Language	People speaking the same language
Social characteristics	Disadvantaged groups, clans, castes
Common heritage	Guest workers, fugitives or involuntarily displaced persons with a common origin
Common practice	Marital arrangements, consanguinity, religious restrictions on abortion
Common problem	Voluntary organizations and support groups, all carriers of a specific mutation

wards international conferences and workshops, with community genetics in their title, have been held in Switzerland (Geneva, 1999), The Netherlands (Amsterdam, 2000), Canada (Jonquière, 2000), Austria (Vienna, 2001), India (Bangalore, 2002), the United Kingdom (Birmingham, 2003) and the United Arab Emirates (Dubai, 2003). In at least one country (The Netherlands), there is a national scientific society for community genetics since 2001.

Community genetics as a practice includes genetic screening, genetic education, monitoring of the population for genetic diseases, interest in quality and access of genetic services, in particular for disadvantaged populations and communities, and promotion of pre-pregnancy consultation and primary care genetics. At a scientific level, the focus is on what a community wants and needs, as well as on the consequences of specified genetic interventions or remissness of interventions for the public. The different aspects of these activities and interests makes the field highly multidisciplinary. Apart from different genetic specialists and non-genetic medical doctors, health scientists, epidemiologists, psychologists, sociologists, ethicists and lawyers among others are involved.

As suggested by its name, the interests of the community are central to community genetics. This goes with a certain weakness for bottom-up initiatives above top-down measures. At least a community consultation is warranted.

In general, communities can be defined at different levels: geographically, linguistically, socially or genetically. Without diminishing the importance of other distinctions, genetic communities are of special interest to our field [15]. Genetic communities can also be defined at different levels: on the basis of common heritage, on the basis of common practice, or on the basis of common problems. Table 2 gives examples of different types of communities. Frequently communities differ from each other at several levels simultaneously.

Conclusion

Community genetics reconciles clinical genetics and public health by combining the best from both fields.

Appendix

Genetic Counseling

Genetic counseling is a communication process which deals with the human problems associated with the occurrence, or the risk of occurrence, of a genetic disorder in a family. This process involves an attempt by one or more appropriately trained persons to help the individual or family to (1) comprehend the medical facts, including the diagnosis, probable course of the disorder, and the available management; (2) appreciate the way heredity contributes to the disorder, and the risk of recurrence in specified relatives; (3) understand the alternatives for dealing with the risk of recurrence; (4) choose the course of action which seems to them appropriate in view of their risk, their family goals, and their ethical and religious standards, and to act in accordance with that decision, and (5) to make the best possible adjustment to the disorder in an affected family member and/or to the risk of recurrence of that disorder. (American Society of Human Genetics, 1975.)

References

1 The New Oxford Dictionary of English. Oxford, Oxford University Press, 2001.
2 American Society of Human Genetics: Ad Hoc Committee on Genetic Counseling. Genetic counseling. Am J Hum Genet 1975;27:240–242.
3 Berkenstadt M, Shiloh S, Barkai G, Katznelson MB, Goldman B: Perceived personal control (PPC): A new concept in measuring outcome of genetic counselling. Am J Med Genet 1999;82:53–59.
4 Michie S, Dormandy E, Marteau TM: The multi-dimensional measure of informed choice: A validation study. Patient Educ Couns 2002;48:87–91.
5 Khoury MJ, Burke W, Thomson EJ (eds): Genetics and Public Health in the 21st Century. Using Genetic Information to Improve Health and Prevent Disease. Oxford Monographs on Medical Genetics. Oxford, Oxford University Press, 2000, vol 40.
6 Juengst ET: 'Prevention' and the goals of genetic medicine. Hum Gene Ther 1995;6:1595–1605.
7 Ten Kate LP: Editorial. Community Genet 2002;5:87.
8 Cuckle H: Extending antenatal screening in the UK to include common monogenic disorders. Community Genet 2001;4:84–86.
9 Lippman A: Bottom line genetics. Commentary to Cuckle. Community Genet 2001;4:87–89.
10 Henneman L, Poppelaars FAM, Ten Kate LP: Evaluation of cystic fibrosis carrier screening programs according to genetic screening criteria. Genet Med 2002;4:241–249.
11 Henneman L, Bramsen I, Van Kempen L, Van Acker MB, Pals G, Van der Horst HE, Adèr HJ, Van der Ploeg HM, Ten Kate LP: Offering preconceptional cystic fibrosis carrier couple screening in the absence of established preconceptional care services. Community Genet 2003;6:5–13.
12 Wald NJ, Morris JK, Rodeck CH, Haddow JE, Palomaki GE: Cystic fibrosis: Selecting the prenatal screening strategy of choice. Prenat Diagn 2003;23:474–478.
13 Modell B: Cystic fibrosis screening and community genetics. J Med Genet 1990;27:475–479.
14 Modell B, Kuliev A: The history of community genetics: The contribution of the haemoglobin disorders. Community Genet 1998;1:3–11.
15 Raz AE, Atar M, Rodnay M, Shoham-Vardi I, Carmi R: Between acculturation and ambivalence: Knowledge of genetics and attitudes towards genetic testing in a consanguineous Bedouin Community. Community Genet 2003;6:88–95.

Community Genetics

Community Genet 2005;8:12–16
DOI: 10.1159/000083331

Genetics and Global Health Care

S. Raeburn

Genetic Unit, College of Medicine and Health Sciences, Sultan Qaboos University, Muscat, Sultanate of Oman

'Imagination is more important than knowledge. Knowledge is limited. Imagination encircles the world.'

Albert Einstein

Key Words
Ataxia telangiectasia · Developing/developed countries · Funding · Genetic epidemiology · Global health care · Gross domestic product · Health care insurance · Retinoblastoma

Abstract
Objectives: To examine an alternative model for funding genetic health care, on a global basis. ***Methods:*** Internet-based national data on gross domestic product (GDP) per capita, health care funding, and public and private elements of health care costs. ***Results:*** Wide variation in GDP per capita and in the proportion available for health care funding. Insufficient funds are available in developing countries. ***Conclusions:*** Health care provision for people with genetic disorders is unlikely to be fully funded unless a different approach to management costs is undertaken. Rare genetic disorders could be funded by an insurance model which may be more equitable and which could be developed to cover the total global health care costs of the genetic disorder.

Introduction

The completion of the Human Genome Project means that we now have a blueprint of human development, with many insights into those gene mutations which can cause significant disease. For some mutations, it is possible to estimate the population-attributable risk and thus to compare their impacts on a community's health status [1–3]. Genetic epidemiologists have already collected information about the most prevalent single gene disorders and some of the variations among populations. The challenge now is to use these data to bring benefit to sufferers from genetic disorders, irrespective of their geographical distribution. Genetic approaches could be valuable in developing a global approach to health care.

Countries that aspire to provide health care must identify funding for all disease management, especially diagnosis, therapy, prophylaxis and social support. Where there is accurate epidemiology, funding can be distributed rationally, and costs to tackle major disease burdens can be estimated and thus fair budgets are provided. However, no state-funded National Health Service (as in the UK) has been able to withstand an inexorable increase in health care costs, even though some disorders have been prevented or minimized. Where health funding is privately provided, costs have also risen significantly. The insur-

KARGER

Fax + 41 61 306 12 34
E-Mail karger@karger.ch
www.karger.com

1422–2795/05/0081–0012$22.00/0

Accessible online at:
www.karger.com/cmg

S. Raeburn
Genetic Unit, College of Medicine and Health Sciences
Sultan Qaboos University, POB 35
Muscat 123 (Sultanate of Oman)
Tel. +968 514451 ext. 3464, Fax +968 513390, E-Mail raeburn@squ.edu.om

ers who provide health care cover have also had to increase premiums.

The shortfall in funding for adequate health care effectively means that people with rarer conditions, such as most single-gene disorders, have been disadvantaged. The infrequency of the disorders means that the pressure on governments to provide specialized services is less (than for common conditions).

This paper summarizes the evidence for examining the world epidemiology of important genetic conditions and for sharing the benefits and costs of health care across national boundaries. Obviously, this would be an equitable solution but it would also add to knowledge by clarifying the factors that influence the presenting phenotypes. It should also help to protect vulnerable families, often living in most adverse conditions. Considerable work has already been reported on the management of hemoglobinopathies internationally [1, 4]. This paper will summarize an approach, which could be applied to rare genetic conditions. My proposal is to pilot it first with two single gene disorders, retinoblastoma and ataxia telangiectasia, representing autosomal dominant and autosomal recessive inheritance, respectively.

Initially, it appears impossible to fund management of these conditions on a worldwide basis or even to achieve agreement on the core health packages. However, there are funding models which could provide for timely relevant interventions and which would contain costs if tackled on a global scale. Prospective studies could begin with modest research resources.

Background

For any government, providing health care for its population is a major challenge; health care is one of the greatest consumers of state resources. Health care systems first evolved with the aim of prevention; subsequently, they became overloaded with the costs of treatment of the major life-threatening conditions and chronic handicaps. Arguably, people with genetic disorders have a greater need for access to quality health care than those without; their state of health depends on timely therapeutic intervention, on pre-emptive prophylaxis and on speedy treatment of intercurrent illness.

The concept of global health care appears to be idealistic, given disparities in the wealth of different nations and even greater cultural and political differences. Yet health care needs, genetic illness by genetic illness, are more similar across international boundaries than are other publicly funded needs. For most genetic diseases, the treatment paradigm is similar, no matter the ethnic group affected or the location. In contrast, provision of total health care globally will be more elusive, given the lack of political will and limited resources of the WHO and other international medical organizations. However, if global care of some genetic diseases were to be effective it could lead to the provision of 'core' medical services for all. Agreements on care standards that should be made available, irrespective of ethnicity or domicile, might attain this.

It is almost inevitable that provision for people with rare genetic disorders will not be fully implemented in state health services. As a Clinical Director of a busy Genetics Directorate in Nottingham, UK, for most of the 1990s, the author observed that access to the proportionately tiny funds (in comparison to other health care costs) required for Genetic Services was often postponed or cancelled to allow funding of major units, e.g. to treat breast cancer, chronic renal disease or ischemic vascular disease. Even the injection of over £50 million in 2003, by a UK government supportive of genetic services [5], could not provide enough to manage people affected by genetic disease and their families. There are so many more common health problems requiring prior access to limited state funds.

An irony is that democratic systems are less likely to provide for people with rare conditions; this is because communities that are offered options will vote for treatments that impinge on their own family and friends most, or for the care of popular groups (e.g. children). Commonest conditions are commonest and so patients with rare problems lose out. Generally, people with rare conditions have insufficient support in the community to provide for their medical needs, which are also more complex and expensive. For rare genetic conditions, special measures are required. This paper sets out an alternative approach to their health care provision. The Discussion incorporates comments and questions which arose at the International Genetic Congress in Dubai (December 9–11, 2003).

Methods

Internet sites were searched for information about the annual gross domestic products (GDP) of six nations or regions [6, 7]. The nations or regions selected for this paper were chosen in order to demonstrate principles and did not include the 50 countries with the lowest GDP per capita. Where available, the proportions of the GDP used for health care, as well as data on insurance provisions for health care, were also collected. The world epidemiology of retinoblastoma and of ataxia telangiectasia was also estimated, helped by information obtained from self-help organizations [8–13]. For many countries these data are incomplete.

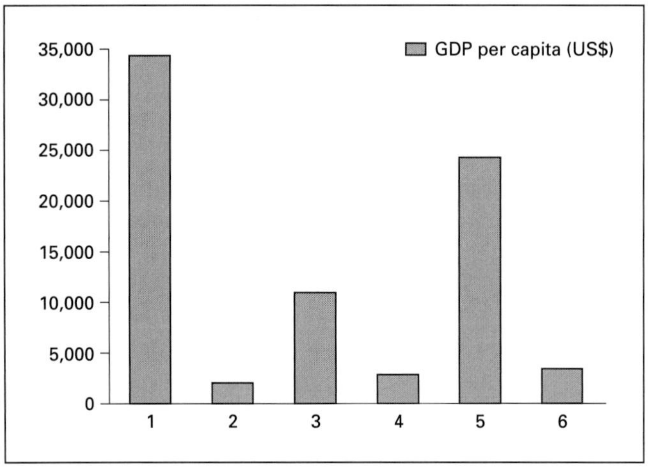

Fig. 1. The wide range of per capita earnings by six world countries/regions during 2001. 1 = USA; 2 = Pakistan; 3 = South Africa; 4 = India; 5 = UK; 6 = Northern Africa and Middle East.

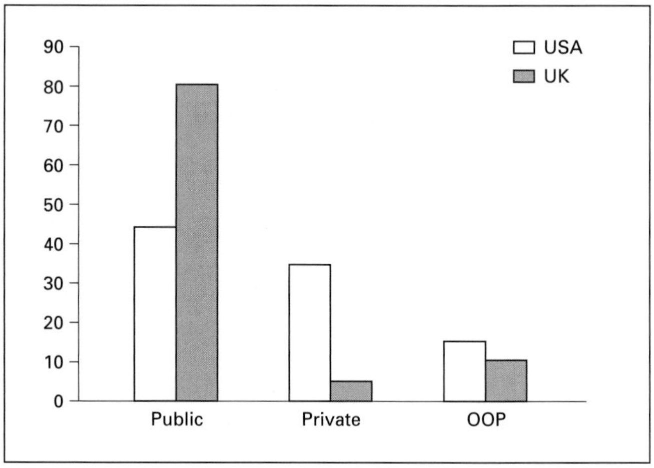

Fig. 2. The proportion of health care costs provided from public funds, private insurances or out of pocket (OOP – payments by the individual) in the USA and the UK.

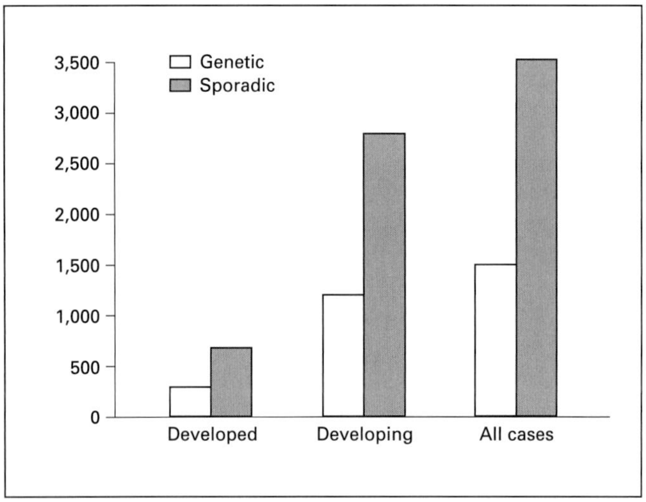

Fig. 3. The annual incidence of retinoblastoma divided into the proportion occurring in developed or developing countries. This figure also shows for each category the proportion of cases expected to be inherited (due to germ line mutations in the retinoblastoma gene).

Results

Figure 1 shows the GDP per capita in dollars for the year 2000 for the USA, Pakistan, South Africa, India, the UK and the Arabic-speaking nations of the Middle East and North Africa. The data are corrected for purchasing power parity. The GDPs vary over a wide range. To provide any health care at all in the poorest countries of the world from their own resources is at best difficult and at worst impossible.

The proportion of the GDP used for health care in these countries also varied (from less than 6 to around 15%, – figures not given here), but not nearly as greatly as the per capita GDP. The greatest variance of funding availability for state-financed health care is in the per capita GDP, not in the percentage that a government apportions.

Figure 2 shows for the USA and the UK how much variation there can be between the percentage of health care paid for from public funds, from private resources (out of pocket) or from insurance. In developing countries, the proportion of out-of-pocket costs used to obtain health care is likely to be as high as in the US because other sources of health care funding may not exist. Those without sufficient money would not receive necessary health care, or only that given charitably.

Figure 3 shows, for the example of retinoblastoma, an initial estimate of the annual incidence in developed and developing nations. Of the estimated 5,000 new cases per annum, 4,000 would occur in developing countries. Retinoblastoma incidence does vary, but for many rare gene mutations the proportion seen in developing/developed countries would be similar to 80/20. Ataxia telangiectasia has differing incidences between countries (data not shown), in part due to differing founder effects; much more variance in incidence is due to differing marriage customs and whether endogamous or consanguineous matings are common [14].

Discussion

There are many barriers to establishing the type of genetic health care outlined in this paper. Greatest of all is the problem of gaining agreement about a disease management from nations in dispute with each other on non-health matters. But a model of funding might be established via an international mechanism, which enabled poor countries to set up core services to diagnose specific disorders; they would then receive assistance to provide the specialist treatment services. There is more similarity between the optimal management of a single gene disorder throughout the world than in treating the complete range of conditions, albeit in one geographical region. Put bluntly, the provision of care limited to one disease would be much more easily achieved than universal health care.

The biggest problem would be funding. It is clearly essential that wealthy countries who contribute now to the health care costs of those with low GDPs decide to coordinate their aid. Political decisions to do this could be linked to other examples of international aid. It is certainly possible that this approach could reduce some of the inequalities that cause conflicts and could thereby lead to defense and policing savings. But is there a long-term model for funding which could be established? The author's view, based on a 7-year period as Genetic Advisor to the Association of British Insurers, is that provision of this type of funding would be feasible using the skills of actuarial assessment and then of evidence-based underwriting, which currently exist internationally in the insurance industry. For example, there could be insurance products which gave a country's health care system cover for the management of, say, retinoblastoma or ataxia telangiectasia. Different insurers would calculate the premiums for such cover in the normal manner, based first on the epidemiology of the disease and the costs that would be expected to occur in a given country. Later, a calculation would be based on the figures collated year by year. The scheme would be similar to group insurance schemes already offered in many countries; the difference is that the group would refer to the country and the risk is of having individuals with retinoblastoma or ataxia telangiectasia (instead of an employment group and the risk of any health problem). Insurers have not explored this approach so far; their interests and resources have been focused on the genetic debate about the insurance underwriter's access to known genetic test results. Would not a model of health care provision that would benefit people in many parts of the world be a more suitable use of insurers' and geneticists' time and skills?

Table 1. The attributes of a genetic condition of importance in researching this approach to health care

1	Sufficiently rare for the overall world health care costs to be manageable
2	Easily diagnosable, even with limited medical resources
3	Specific genetic confirmation of the diagnosis possible for exact classification
4	Specialist treatments available with a high probability of altering the prognosis
5	Follow-up methods applicable in each country with a system for ongoing audit linked with 2 above

The choice of conditions to study in the health care scheme proposed here should not be made by one geneticist or one group of people with genetic disadvantages. However the first attributes of a disease to study would be as in table 1. Retinoblastoma and ataxia telangiectasia could perhaps be examples, but many more discussions and incidence studies are required. If the principle of global health care provision for some rare conditions is to be explored, the discussions, which concern which genetic diseases to study, will in themselves be of great value.

Acknowledgments

The author wishes to thank Prof. Bernadette Modell and several colleagues at the College of Medicine and Health Sciences of the Sultan Qaboos University, Sultanate of Oman, for extremely valuable discussions, as well as for the patience to consider a radical and unusual suggestion in depth.

References

1 Alwan A, Modell B: Recommendations for introducing genetics services in developing countries. Nat Rev Genet 2003;4:61–68.
2 Warren V, Brett P, Brayne, C, Macdonald AS, Plumb RH, Read AP: Genetic tests and future need for long-term care in the UK. Report of a Work Group of the Continuing Care Conference Genetic Tests and Long-Term Care Study Group. London, Continuing Care Conference, 1999.
3 World Health Report 2002, WHO, Geneva.
4 World Health Organization: Guidelines for control of haemoglobin disorders. Geneva, World Health Organization, 1994 (unpubl document WHO/HDP/HB/GL/94.1).
5 Department of Health: Genetics White Paper: Our Inheritance, Our Future – Realising the Potential of Genetics in the NHS. Norwich, Stationery Office, 2003, CM 5791, para. 5.21.
6 Population Reference Bureau. 2003. http://www.prb.org/datafind/datafinder.htm. January 2004.
7 OECD Health Data, ed 3. Tables 11, 12, 16, 17. http://www.oecd.org/document/16/0,2340,en_2825_495642_2085200_1_1_1_1,00.html.
8 Yeole BB, Advani SH: Retinoblastoma: An epidemiological appraisal with reference to a population in Mumbai, India. Asian Pacific J Cancer Prev 2002;3:17–21.
9 Draper GJ, Sanders BM, Brownbill PA, Hawkins MM: Patterns of risk of hereditary retinoblastoma and applications to genetic counselling. Br J Cancer 1992;66:211–219.
10 Lohmann D, Scheffer H, Gallie B: Best Practice Guidelines for Molecular Analysis of Retinoblastoma. European Molecular Genetics Quality Network, EMQN, 2002. http://www.emqn.org/Assets/uploadpdfs/RB.pdf.
11 Noorani HZ, Khan HN, Gallie BL, Detsky AS: Cost comparison of molecular versus conventional screening of relatives at risk for retinoblastoma. Am J Hum Genet 1996;59:301–307.
12 Swift M, Morrell D, Cromartie E, Chamberlin AR, Skolnick MH, Bishop DT: The incidence and gene frequency of ataxia telangiectasia in the United States. Am J Hum Genet 1986;39:573–583.
13 Stankovic T, Kidd AM, Sutcliffe A, McGuire GM, Robinson P, Weber P, Bedenham T, Bradwell AR, Easton DF, Lennox GG, Haites N, Byrd PJ, Taylor AMR: ATM mutations and phenotypes in ataxia-telangiectasia families in the British Isles: Expression of mutant ATM and the risk of leukemia, lymphoma and breast cancer. Am J Hum Genet 1998;62:334–345.
14 Bittles AH: Consanguinity and its relevance to clinical genetics. Clin Genet 2001;60:89–98.

Community Genetics

Community Genet 2005;8:17–20
DOI: 10.1159/000083332

Endogamy, Consanguinity and Community Disease Profiles

A.H. Bittles

Centre for Human Genetics, Edith Cowan University, Perth, Australia

Key Words
Consanguinity · Endogamy · Homozygosity · Wahlund effect

Abstract
Considerable attention is paid to the role of consanguineous marriage as a causative factor in the prevalence of genetic disorders. At the same time, the potential influence of community endogamy on overall levels of homozygosity and disease profiles remains largely underinvestigated. With the ongoing global epidemiological transition from infectious to non-communicable disease, the impact of genetic disorders will become increasingly important and a thorough understanding of the determinants of human population-genetic structure will be all the more necessary. In particular, the genetic components of adult-onset diseases will become more obvious and assume greater significance. Similarly, refinements of study design to incorporate intercommunity genetic variation appear to be an essential prerequisite in pharmacogenetic research if the concept of individualized treatments is to achieve reality, with equivalent subject-control comparison difficulties also predicted in forensic genetics.

Introduction

With the mapping of the human genome attention has changed to the topic of human genetic variation, and in particular to the specific disease profiles that can occur in different populations. The topics of race and ethnicity, and whether they should be considered in health studies and clinical practice, attract diametrically opposed opinions [1, 2]. Whichever stance is adopted, microsatellite studies at autosomal loci have shown that 93–95% of genetic variation in humans is within-population, with allelic differences between major ethnic populations accounting for just 3–5% of total variation [3]. From a clinical perspective, this suggests that greater attention needs to be paid to the nature and extent of disease variation between different subpopulations. The aims of this paper are to examine how these subdivisions could have arisen, the extent to which they are maintained, and their effects on community health.

The Influence of Endogamous Marriage on the Gene Pool

The rapid expansion in numbers of humans over the last three centuries has been accompanied by increasing migration of individuals and communities across and between continents. In pre-industrial societies most marriages occurred within quite narrow geographical boundaries, which limited gene transmission between commu-

KARGER
Fax +41 61 306 12 34
E-Mail karger@karger.ch
www.karger.com

Accessible online at:
www.karger.com/cmg

Prof. A.H. Bittles
Centre for Human Genetics, Edith Cowan University
100 Joondalup Drive
Perth, WA 6027 (Australia)
Tel. +61 8 6304 5623, Fax +61 8 6304 5851, E-Mail a.bittles@ecu.edu.au

nities. Genetic admixture was further restricted by positive assortative mating, i.e. by individuals choosing a marriage partner not only from the same geographical area but also with similar physical characteristics, religious beliefs, educational backgrounds, and even political opinions. Through time, whether via self-selection on the part of the couple or marriage arrangements contracted by their parents, assortative mating predictably would have resulted in genotypic differences between communities and populations.

Although favored in many populations, the practice of endogamous marriage, i.e. marital unions within defined communities, has attracted less attention. To some extent this reflects the preferences and practices of modern western societies, in which marriage is held to be a matter of personal choice, and the community or ethnic background of a potential partner may be considered of limited relevance. Where there has been a move from intra-community marriage the change has been accompanied by the mixing of previously separated gene pools, resulting in increased overall genetic heterogeneity.

Marriage within clan, tribal, caste or *biraderi* boundaries largely remains the rule in more traditional societies, and endogamous marriage also is strongly favored in many migrant communities now resident in western countries. In these subpopulations there has been limited opportunity for genetic admixture and so greater intra-community genetic homogeneity would be expected. A number of factors mediate the influence of endogamy on community gene pools, such as the number of founders, the age of the community, community size, the flexibility or rigidity of its marriage boundaries, and population amalgamations through time, including religious conversions. While each of these variables could influence the structure of all endogamous populations, they are essentially community specific in their action and hence in their impact on gene pool composition. However, the net result would be an increased proportion of homozygotes in the population as a whole, i.e. the Wahlund effect, that could be mistakenly interpreted as the outcome of consanguineous marriage [4].

The Prevalence and Distribution of Consanguineous Marriage

Consanguinity is generally uncommon in Europe, North America and Australasia, in part because of questionable scientific and medical opinion dating from the mid 19th century [5]. By comparison, consanguineous marriage is quite prevalent in South America and among many African tribes, and until recently marriages between first cousins with different surnames were permitted in the majority Han population of China. However, it is in North Africa, the Middle East, and much of Central and South Asia that the highest rates of consanguineous marriage are observed, with some 20 to over 50% of all unions between couples related as second cousins or closer [6].

Besides major religious differentials, local and regional societal regulations, concepts and beliefs can influence the prevalence and specific types of consanguineous marriage contracted. Although the *Quran* contains no specific guidance encouraging consanguinity [7], as in most Muslim countries, marriage between first cousins is a long-standing tradition in the Gulf States. With community endogamy and a widespread preference for consanguineous unions, each community has in common a higher level of autozygosity, which in turn would be expected to result in a greater prevalence of recessive disorders. There may, however, be major differences in the profiles of deleterious mutations within and between specific communities and, on a larger scale, between populations.

The excess risk that an autosomal recessive disorder will be expressed in the progeny of consanguineous unions is inversely proportional to the frequency of the disease allele(s) in the gene pool. For this reason, during the last decade many rare disease genes have been identified, and their chromosomal locations mapped, by studying genetically isolated populations and highly inbred families with multiple affected members. As has been demonstrated in the populations of the Gulf region, these disorders frequently are specific to individual clans or tribes [8]. Besides consanguinity, given their often small effective population sizes, it would be expected that founder effect and random drift also operate and influence the prevalence of inherited disorders.

The Clinical Implications of Consanguinity and Variant Community Disease Profiles

A multinational study of over 600,000 pregnancies and live births showed that prereproductive mortality was 4.4% higher in first cousin progeny ($F = 0.0625$) than in the offspring of non-consanguineous unions [9]. Using data on the mean coefficient of inbreeding (α) in specific countries (http://www.consang.net), this estimate can be used to determine the potential influence of consanguinity on early mortality in different populations of the Gulf region, and in countries which provide migrant workers to the region (table 1).

A number of important qualifications need to be made in the interpretation of these results. Most of the studies cited were conducted in developing countries, with high background levels of infant and under-5-year mortality, and consanguinity-associated deaths largely concentrated during the 1st year of life. Due to poor diagnostic facilities a specific cause of death was seldom identified, and so the actual contribution of genetic disorders to premature mortality was unclear. There also was limited control for non-genetic variables that can influence early survival, a situation that was investigated in the 1990–1991 Pakistan Demographic and Health Survey using multivariate logistic regression. As indicated in table 2, analysis of these data showed that non-genetic variables such as restricted birth interval, maternal illiteracy and young maternal age exerted a greater adverse effect than consanguinity ($F = 0.0625$) on deaths in infancy and from 2 to 5 years of age [10].

Diagnostic and investigative limitations do not apply in most Gulf States, and as a result many previously unreported genetic disorders have been identified, especially in inbred kindred. The rapid transition from high infant and childhood mortality to extended morbidity has also meant that there is the genuine prospect of survival of affected individuals into adulthood [11]. As previously predicted, this shift in disease profile may influence community attitudes to consanguineous marriage [6], but it also can create difficulties for couples who, despite knowing the cause of their progeny's disease, may feel obliged to fulfill their parental obligations by proceeding with marriage arrangements for their affected offspring. A decision of this type could be further complicated in communities where intrafamilial marriage is preferential, since the probability of disease transmission to any children that may ensue will be high. However, it also has to be recognized that marriage with non-relatives may become virtually impossible once the incapacity of the affected individual is known to the wider community.

Table 1. Consanguinity and prereproductive mortality in selected countries

Country	Mean coefficient of inbreeding (α)	Estimated number of consanguinity-associated deaths
Bahrain	0.0159	11.2/1,000
Egypt	0.0230	16.2
Iran	0.0131	9.2
Iraq	0.0235	16.5
Jordan	0.0246	17.3
Lebanon	0.0091	6.4
Oman	0.0198	13.9
Pakistan	0.0278	19.4
Palestine	0.0198	13.9
Saudi Arabia	0.0208	14.6
United Arab Emirates	0.0245	17.2

Table 2. Major determinants of infant and 2- to 5-year mortality in Pakistan

Rank order	Determinants	OR
Infant mortality		
1	Birth interval <18 months	2.70
2	Maternal illiteracy	2.42
3	Maternal age <20 years	1.52
4	First cousin offspring	1.32
2- to 5-year mortality		
1	Birth interval <18 months	5.31
2	Maternal illiteracy	1.83
3	Maternal age <20 years	1.56
4	Male sex	1.48

No significant increase was observed in 2- to 5-year mortality among first cousin offspring. OR = Odds ratio.

Discussion

Although much attention has focused on the role of consanguinity in the prevalence of recessive disorders, it is endogamy which principally governs the spectrum of observed diseases in a community and, as previously noted, a higher overall level of homozygosity is observed in genetically subdivided populations. The greater the degree of genetic isolation and the longer its history of separation, the more probable that unique disorders will have arisen in a community. For this reason, the presence of multiple discrete subpopulations creates substantial practical difficulties in the compilation of disease registers and the organization of screening programs for genetic disorders at national and even regional levels [12].

Once a mutation has been identified within a consanguineous pedigree, case ascertainment in other family members can usually be greatly simplified and genetic counselling facilitated [13]. Caution is, however, merited since an analysis of β-thalassaemia mutations in Pakistan found that 11.7% of first cousins with the disease were double heterozygotes [14], indicating a high prevalence of different thalassaemia mutations in the population and the probability of significant past gene admixture [15].

Endogamy and preferential consanguineous marriage have significant practical implications for developing disciplines such as pharmacogenetics, since the choice of appropriate study and control populations will pose major problems in a country with multiple discrete subcommunities, and a treatment that is successful in one community may be clinically ineffective in others. Similar problems can arise in DNA-based forensic studies, as many individuals from a population isolate may share a high proportion of alleles across loci, with multiple consanguineous pathways in individual kindred. A situation of this type can create major difficulties in the calculation of forensic match probabilities [4, 16, 17].

Given the rapid extension of life expectancy in developed and developing countries, it is important that attention be paid to the influence of endogamy and consanguinity on adult-onset diseases, a topic which to date has been largely ignored. Once again, it must be appreciated that in endogamous societies, genetic predisposition to specific diseases may be community-specific [6, 12], which could explain at least some of the apparently contradictory findings that have been obtained in breast cancer studies conducted in different populations [18, 19]. Without protocols specifically designed to account for the contribution of community-specific mutations, discussion on this and equivalent subjects will remain speculative.

References

1 Cooper RS, Kaufman JS, Ward R: Race and genomics. Science 2003;348:1166–1170.
2 Gonsález Burchard E, Ziv E, Coyle N, Gomez SL, Tan H, Kartner AJ, Mountain JL, Pérez-Stable EJ, Sheppard D, Risch N: The importance of race and ethnic background in biomedical research and clinical practice. Science 2003;348:1170–1175.
3 Rosenberg NA, Pritchard JK, Weber JL, Cann HM, Kidd KK, Zhivotovsky LA, Feldman MW: Genetic structure of human populations. Science 2002;298:2381–2385.
4 Overall ADJ, Nichols RA: A method for distinguishing consanguinity and population substructure using multilocus genotype data. Mol Biol Evol 2001;18:2048–2056.
5 Bittles AH: The bases of western attitudes to consanguineous marriage. Dev Med Child Neurol 2003;45:135–138.
6 Bittles AH: Consanguinity and its relevance to clinical genetics. Clin Genet 2001;60:89–98.
7 Hussain R: Community perceptions of reasons for preference for consanguineous marriages in Pakistan. J Biosoc Sci 1999;31:449–461.
8 Teebi AS, Farag TI (eds): Genetic Disorders among Arab Populations. New York, Oxford University Press, 1997.
9 Bittles AH, Neel JV: The costs of human inbreeding and their implications for variations at the DNA level. Nat Genet 1994;8:117–121.
10 Grant JC, Bittles AH: The comparative role of consanguinity in infant and childhood mortality in Pakistan. Ann Hum Genet 1997;61:143–149.
11 Alwan A, Modell B: Recommendations for introducing genetics services in developing countries. Nat Rev Genet 2003;4:1–8.
12 Bittles AH: Consanguineous marriage and childhood health. Dev Med Child Neurol 2003;45:571–576.
13 Ahmed S, Saleem M, Modell B, Petrou M: Screening extended families for genetic hemoglobin disorders in Pakistan. N Engl J Med 2002;347:1162–1168.
14 Ahmed S, Petrou M, Saleem M: Molecular genetics of β-thalassaemia in Pakistan: A basis for prenatal diagnosis. Br J Haematol 1996;94:476–482.
15 Wang W, Sullivan SG, Ahmed S, Chandler D, Zhivotovsky LA, Bittles AH: A genome-based study of consanguinity in three co-resident endogamous Pakistan communities. Ann Hum Genet 2000;64:41–49.
16 Mohyuddin A, Ayub Q, Qamar R, Zerjal T, Helgason A, Mehdi SQ, Tyler-Smith C: Y-chromosomal STR haplotypes in Pakistani populations. Forensic Sci Int 2001;118:141–146.
17 Zhivotovsky LA, Ahmed S, Wang W, Bittles AH: The forensic DNA implications of genetic differentiation between endogamous communities. Forensic Sci Int 2001;119:269–272.
18 Denic S, Bener A: Consanguinity decreases risk of breast cancer – Cervical cancer unaffected. Br J Cancer 2001;85:1675–1679.
19 Liede A, Malik IA, Aziz TZ, de los Rios P, Kwan E, Narod SA: Contribution of BRCA1 and BRCA2 mutations to breast and ovarian cancer in Pakistan. Am J Hum Genet 2002;7:595–606.

Community Genetics

Community Genet 2005;8:21–26
DOI: 10.1159/000083333

Genetic Diversity among the Arabs

Ahmad S. Teebi[a, c] Saeed A. Teebi[b]

[a]Section of Clinical Genetics and Dysmorphology, Division of Clinical and Metabolic Genetics, and
[b]Center for Computational Biology, Hospital for Sick Children, Toronto, Canada;
[c]Department of Medical Genetics, King Faisal Specialist Hospital and Research Center, Riyadh, Saudi Arabia

Key Words
Autosomal recessive disorders · Bedouin · Consanguinity · Heterogeneity · Homozygosity · Isolates · Middle East

Abstract
The Arabs in general are genetically diverse. Major factors that contributed to their diversity include the migrations of Semitic tribes from the Arabian Peninsula, the Islamic expansion in the 7th century AD, the Crusade wars and the recent migration dynamics. These events have resulted in the admixture of the original Arabs with other populations extending from east and south Asia to Europe and Africa. Their demographic features include high rates of consanguinity, a large family size and a rapid population growth. There is a high frequency of autosomal recessive disorders and increased frequencies of homozygosity for autosomal dominant traits, such as familial hypercholesterolemia and X-linked traits, such as glucose-6-phosphate dehydrogenase deficiency. The patterns of autosomal recessive disorders, including their mutations, may be different in various geographic locations within the Arab world. However, there are disorders that are specifically prevalent among the Arabs either uniformly or in certain locations. The Arab Genetic diseases include Bardet-Biedl syndrome, Meckel syndrome, autosomal recessive severe childhood muscular dystrophy, osteopetrosis and renal tubular acidosis, Sanjad-Sakati syndrome and others.

Introduction

The interest in the study of genetic disorders among the Arabs is relatively new. While a significant amount of information has originated from some Arab countries such as Lebanon, Tunisia, Saudi Arabia, Egypt, Jordan and the occupied Palestinian territories, Kuwait and the United Arab Emirates (UAE), as well as from the study of Arab immigrants in the Western World, the information is still far from being sufficient.

The publication of the book by Teebi and Farag [1] has directed attention to the problem and contributed to the increasing interest in this field. This was followed by a number of conferences by the Middle East Genetic Association of America in Tunisia [2], Egypt [3] and Saudi Arabia, and meetings organized and sponsored by the WHO, the European Community and some Arab countries. The recent establishment of the Arab Genetic Disease Database (www.agddb.org) [4] as well as the efforts by the Arab Genome Center in UAE to create other databases adds another dimension towards the progress in this field in the Arab world by providing more up-to-date sources of information.

Arab People

The Arabs are united by a history that dates back to 3500 BC. They originated from Semitic tribes in the Arabian Peninsula spreading in the Valley of Tigris and Eu-

KARGER

Fax +41 61 306 12 34
E-Mail karger@karger.ch
www.karger.com

1422–2795/05/0081–0021$22.00/0

Accessible online at:
www.karger.com/cmg

Prof. Ahmad S. Teebi
Division of Clinical and Metabolic Genetics
Hospital for Sick Children, 555 University Avenue
Toronto, Ontario M5G 1X8 (Canada)
Tel. +1 416 813 8366, Fax +1 416 813 5345, E-Mail ateebi@sickkids.ca

phrates rivers initially, and later on (2500 BC) in the East Mediterranean region, which contributed greatly to their diversity. Their diversity was enriched by the Islamic expansion in the 7th century AD, the Crusades and the recent migration dynamics. The Arabs are united by a common language, and their location extends from the Atlantic Ocean in the west, to the Arabian Gulf in the east, spanning an area of about 14 million km^2, and is inhabited by a population of 300 million, the majority of which are Muslims.

In addition to Muslim and Christian Arabs, the area is home to Kurds, Berbers, Armenians, Circassians, Jews and other minorities. The Arabs themselves, in most parts of the Arab world, are the result of admixture with neighboring populations such as Persians, Turks and Africans, and distant populations such as Southeast Asians and Europeans. In addition to this mixture, relatively homogeneous populations, or isolates, also exist. These include some Bedouin tribes, Nubians and Druze among others.

The Arab Family and Consanguinity

The Arabs in general have a large family size with four or more children per family. In most Arab countries, a married man and his wife often live with the man's parents in one large household (or at least very close to him) whenever possible, and marriages between relatives are favored. Consanguineous marriages among the Arabs, unlike in Western populations, are not necessarily limited to geographic or religious isolates or ethnic minorities. Although it is not a religious prescription, it is deeply rooted in the culture and has been practiced over many generations. The consanguinity rates are generally high, ranging between 25% in Lebanon [5], up to 60% in Saudi Arabia [6] and 90% in some Bedouin tribes. The rates are still high in Arab immigrants in the West. An average figure of about 40% appears to hold for most Arab countries. The most common form of intermarriage is between first cousins, particularly paternal first cousins, and includes double first cousin marriages. Uncle/niece and aunt/nephew marriages are strictly forbidden in Islam and are actually nonexistent. Consanguineous marriages are more common among Muslims than among Christians and are more common in rural areas than urban regions. It is to be noted that in most Arab societies the rate of inbreeding is not declining [7].

Profile of Genetic Disorders among the Arabs

The increased rates of consanguineous marriages have contributed to the increased frequencies of genetic disorders, particularly the autosomal recessive conditions [8]. Furthermore, this has resulted in increased frequencies of homozygosity for autosomal dominant disorders such as familial hypercholesterolemia in the Lebanese and Bedouin as well as homozygosity for X-linked disorders, such as glucose-6-phosphate dehydrogenase in Oman [1]. In addition, there is an apparently high number of previously unrecognized autosomal recessive syndromes being described from the Arab populations.

Autosomal Recessive Disorders

Arabs have a recognizable pattern of autosomal recessive disorders which is briefly discussed below.

Hemoglobinopathies

The genes for sickle cell hemoglobin (HbS) as well as α- and β-thalassemias are encountered in all Arab countries although significant differences exist from one country to the other and sometimes even within the same country. HbS trait frequency ranges from less than 1% along the Nile in Egypt to 20% or more in Bahrain and the Eastern Province of Saudi Arabia. HbC is very rare except in Morocco, where it is twice as common as HbS. A number of new and extremely rare variants have been reported, an example being Hb0-Arab. Molecular studies of HbS indicate an independent mutation (Arab-Indian haplotype) that occurred in eastern Saudi Arabia and the Indian subcontinent, spreading from there to other parts of the Arab world; while the genes in Western Saudi Arabia and North Africa are believed to have been conveyed from Benin and Senegal in Africa [1, 9]. No specific β-thalassemia mutations are confined to the Arabs, and the Mediterranean and Asian mutations are encountered at variable frequencies reflecting the geographical and historical background of each region [10].

Familial Mediterranean Fever

This condition is very common in Lebanon, Jordan and among Palestinians while being nonexistent in the Arabian Peninsula. The frequency of the disease among Jordanians and Palestinians has been estimated to be around 1:1,500, which is comparable to the prevalence among Armenians and Sephardic Jews [1]. Five mutations in the MEFV gene account for the majority of patients with familial Mediterranean fever in the Middle East, two of which are more common among Arabs [11, 12]. Most patients with familial Mediterranean fever come from a very ancient pool of founders, which suggests common origins for several Middle Eastern populations.

Neuromuscular Disorders

Common neuromuscular disorders encountered in the Arab population include spinal muscular atrophy with a prevalence of 1:1,000 live births in some areas; severe childhood autosomal recessive muscular dystrophy with a prevalence similar or slightly less than that of the Duchenne type particularly in Sudan, Tunisia, and Kuwait; congenital muscular dystrophy; peripheral neuropathies; cerebellar ataxia, and hereditary spastic paraplegia [1, 13].

Metabolic Disorders

Due to the lack of metabolic diagnostic facilities and nationwide screening programs in the majority of Arab counties, available data are mostly derived from Saudi Arabia, Kuwait, and Israel. In Saudi Arabia, high frequencies of organic acidemias have been reported and these include propionic acidemia, methylmalonic acidemia, isovaleric acidemia and HMG-CoA-lyase deficiency, with a clear trend for tribal predilection. Homocystinuria is common in Qatar and Kuwait. Other frequently diagnosed disorders among the Arabs include maple syrup urine disease, argininosuccinic aciduria, nonketotic hyperglycinemia, tyrosinemia type 1, cystinuria, primary hyperoxaluria and vitamin-D-dependent rickets.

Classic phenylketonuria (PKU) and hyperphenylalaninemia are relatively well-studied entities. The PKU patients constituted 1.6–2.3% of mentally retarded patients studied in Kuwait and Egypt [1], the figures being similar to those in the West in the era before the introduction of neonatal screening. The prevalence of classic PKU in the heterogeneous Arab population of Kuwait has been estimated at 1:6,500 live births [1]. In a nearby geographic region, the Eastern Province of Saudi Arabia, no cases of PKU were detected among 70,000 newborns screened [1]. At another location in Saudi Arabia (Riyadh), patients with hyperphenylalaninemia secondary to 6-pyruvoyltetrahydropterin synthase deficiency were frequently seen and appeared to be more common than those with classic PKU. Mutations and polymorphisms at the phenylalanine hydroxylase gene in several Arab populations, including Palestinian, Egyptian, Kuwaiti and North African populations, have shown a mixture of common European, African and unique mutations.

Lysosomal storage disorders constitute a large part. Among the commonly diagnosed conditions are mucopolysaccharidose syndromes, multiple sulfatase deficiency, Niemann-Pick A, B, and C, GM_2 gangliosidoses, Canavan disease, metachromatic leukodystrophy, Krabbe disease, Gaucher disease (neuropathic type) and neuraminidase deficiency. Some of these disorders have a tribal predilection, the examples being Canavan disease and Niemann-Pick disease type C [1, 8]. Metachromatic leukodystrophy and Krabbe disease are particularly more frequent among the Palestinians in restricted areas.

Genodermatosis

The frequently encountered disorders include cutis laxa, wrinkly skin syndrome, various types of epidermolysis bullosa, Papillon-Lefevre syndrome, xeroderma pigmentosum, Mal de Meleda, ectodermal dysplasias, gerodermia osteodysplastica, Sjögren-Larsson syndrome, and lamellar ichthyosis among others.

Osteopetrosis Syndromes

Data from several countries within the Arabian Peninsula have indicated that the severe autosomal recessive form of osteopetrosis is relatively common [1]. Osteopetrosis with renal tubular acidosis due to carbonic anhydrase II deficiency has been frequently reported from the Arabian Peninsula and Maghreb countries. The majority of hitherto reported patients are Arabs, and the majority of them are homozygous for a unique mutation that appears to be confined to Arabs [1].

Chondrodystrophies

Various autosomal recessive dwarfing syndromes are frequently diagnosed among Arab populations. These include spondyloepiphyseal dysplasia congenita and tarda, spondylocostal dysplasia, pseudorheumatoid arthropathy, Wolcott-Rallison syndrome, fibrochondrogenesis and other lethal chondrodystrophies, and a number of unique forms of bone dysplasias [1, 8, 14]. The Stüve-Wiedemann syndrome has been found to be prevalent in the UAE [15] while the Dyggve-Melchior-Clausen disease is suggested to have a high frequency in the Lebanese and Palestinians.

Cystic Fibrosis

Cystic fibrosis among the Arabs is probably as common as it is in the Caucasian populations in the West. However, the diagnosis is often missed either because of mild manifestations or atypical presentation [1]. It is estimated that the prevalence ranges from 1:2,200 to

1:4,200 live births [1]. In some areas, it has been found to be unusually high: in Qatar, for example, the prevalence was found to be 1:600 live births in a Qatari tribe, with all cases being homozygous for a unique mutation [16]. The prevalence was noted to be much higher in a Palestinian village reflecting founder effect.

Molecular studies from several Arab populations have shown that the F508del mutation accounts for 22–25% of Palestinians and ranges between 15 and 50% of cystic fibrosis chromosomes in other Arab populations. This discrepancy is probably related to Palestinians' proximity to Europe and their history of admixture with European populations. Other European CFTR mutations have also been detected in addition to the unique mutations encountered only among specific Arab populations [17–19].

Congenital Chloride Diarrhea

Its prevalence has been estimated to range from 1: 5,500 live births in Saudi Arabia to 1:13,000 in Kuwait [1]. The condition is probably extremely rare in other Arab countries.

Malformation Syndromes

Two malformation syndromes are common among the Arabs, namely the Bardet-Biedl syndrome and the Meckel syndrome. The Bardet-Biedl syndrome is frequently diagnosed in most Arab countries and is particularly common among the Bedouins of Kuwait, with an estimated prevalence of 1:6,900 live births [1]. The condition was studied at the molecular level in Israel using the Bedouin from Negev desert, and at least eight loci were identified [20]. The Meckel syndrome was found to have a prevalence of 1:2,000 to 1:3,500 live births from Jerusalem and Kuwait, respectively [1]. Also, it is frequently diagnosed in Egypt and the Arabian Peninsula.

Other relatively common syndromes include multiple pterygium syndrome, lethal multiple pterygium syndrome, Bartsocas-Papas syndrome, the persistent Müllerian duct syndrome and autosomal recessive hydrocephalus [1, 8].

Nonsyndromic Deafness

In several Palestinian villages as well as some areas from Tunisia and Oman, nonsyndromic autosomal recessive deafness has been found to be very common. While connexin 26 mutations account for about 30–50% of all cases [21], mitochondrial mutations and other gene mutations have also been found.

Other Disorders

A number of other disorders appear to have special importance in some Arab populations and deserve to be mentioned. They include for example nesidioblastosis of the pancreas in the Arabian Peninsula, cytochrome b5-reductase deficiency in South Algeria, vitamin-E-deficient cerebellar ataxia in Tunisia, hypophosphatemic rickets with hypercalcuria in a large Beouin tribe in Israel, arthrogryposis multiplex congenita-neurogenic types among Palestinian isolates, cerebrotendinous xanthomatosis among Druze in Israel, and Glanzmann thrombasthenia in Jordan. This is in addition to the frequently diagnosed apple-peel jejunal atresias, congenital hypoplastic anemia of Blackfan and Diamond and congenital hepatic fibrosis among others [1, 8].

New Syndromes and Variants

During two decades (1975–1995), the description of new genetic syndromes constituted a 'hot point' subject for research and publication. At least 125 syndromes were first recognized among the 'previously poorly studied' Arab population, with the majority being autosomal recessive [1]. In 2004, the number of new genetic syndromes among the Arabs has exceeded 200 [unpublished]. Most reports came from Lebanon, UAE, Saudi Arabia, Kuwait, Tunisia, Israel and from North America and Europe reporting on Arab families.

Among the well-characterized disorders are the limb/pelvis hypoplasia/aplasia syndrome (Al-Awadi syndrome), the Sanjad-Sakati syndrome and the autosomal dominant Teebi hypertelorism syndrome.

A new type of Ehlers-Danlos syndrome with tortuous blood vessels has recently been characterized from a large tribe from Saudi Arabia and Qatar [22].

Homozygosity for Two or More Recessive Traits

As a consequence of inbreeding, it is not uncommon to observe more than two autosomal recessive conditions in the same sibship. Many examples have been recorded in Arabs [1, 8]. Also, there are examples of families with children having more than one disorder with different modes of inheritance.

Table 1. Genetic diseases that are highly prevalent among the Arabs

Diseases	OMIM No.	Estimated frequency/live births
Bardet-Biedl syndrome	209900	1–2/13,000
Congenital chloride diarrhea	214700	1/5,500
Sanjad-Sakati syndrome	241410	Unknown; all reported patients are Arabs
Meckel syndrome	249000	1/3,500
Spinal muscular atrophy type I	253300	1–2/1,000
Severe childhood autosomal recessive muscular dystrophy	253700	1/3,500 (approximate)
Nesidioblastosis of the pancreas	256450	1/2,675
Osteopetrosis with renal tubular acidosis	259730	Unknown; 70% of all reported patients are Arabs

OMIM = Online Mendelian Inheritance in Man.

Arab Genetic Diseases

Table 1 includes genetic diseases that are characteristically highly prevalent among the Arabs (adapted from Teebi and Farag [1].

Arab Genetic Disease Database

The Arab Genetic Disease Database (AGDDB) is a curated catalogue of genetic disorders found in Arab populations that was established in 2001 [4] by a group in Toronto, headed by Dr. A.S. Teebi. AGDDB contains information on the clinical, genomic, protein, population and reference aspects of Arab disorders (published and otherwise), along with extensive links to entries in OMIM, PubMed and other databases. Updating of the database is carried out via the Arab Genetic Disease Consortium (30 investigators, 18 countries), which is responsible for editing and reviewing AGDDB data. AGDDB is fully searchable and freely available and may be accessed at http://www.agddb.org/.

In addition to this, efforts have been made to create similar databases by the Center for Arab Genomic Studies in the UAE.

Comment

There is a real problem in the Arab world regarding the magnitude of genetic disorders and the factors contributing to their higher frequencies. At the same time, the genetic services are scant and those who work in this field are scarce. While the priorities of Arab governments are to fight communicable disorders, eradicate endemic diseases, and to improve the nutritional status of their people and their basic health standards, it remains of utmost importance to work to reduce the load of genetic disorders, many of which are fatal or constitute a heavy burden on health and social services because of the nature of non-curable problems and handicaps. This work should be done in part by charitable organizations and private sectors or by those who feel that the humanitarian and scientific merits of such work in this area would satisfy their ambition and outlook for healthy generations.

Acknowledgments

We thank Dr. N. Sakati for fruitful discussions, Dr. Al-Owain and Ms. S. Kennedy for reading and Ms. M. Mendoza for secretarial assistance.

References

1 Teebi AS, Farag TI (eds): Genetic Disorders among Arab Populations. New York, Oxford University Press, 1997, p 499.
2 Fathallah D, Dellagi Q, Qumsiyeh M, Teebi AS: Recent advances in molecular genetics of the Maghreb and the Middle East populations: First MEGA Conference, Tunisia. Am J Genet 1998;78:90–91.
3 Teebi AS, Shawky RM: Population and molecular genetic update: The second Middle East Genetic Association of America (MEGA) conference. Am J Med Genet 2000;92:293–294.
4 Teebi AS, Teebi SA, Porter CJ, Cuticchia J: Arab genetic disease database (AGDDB): A population-specific clinical and mutation database. Hum Mutat 2002;19:615–621.
5 Khlat M, Khudr A: Cousin marriages in Beirut, Lebanon: Is pattern changing? J Biosoc Sci 1984;16:369–373.
6 El-Hazmi MA, Al-Swailem AR, Warsy AS, Al-Swailem AM, Sulaiman A, Al-Meshari AA: Consanguinity among Saudi Arabian population. J Med Genet 1995;32:623–626.
7 Hoodfar E, Teebi AS: Genetic referrals of Middle Eastern origin in a western city: Inbreeding and disease profile. J Med Genet 1996;33:212–215.
8 Teebi AS: Autosomal recessive disorders among Arabs: An overview from Kuwait. J Med Genet 1994;33:224–233.
9 El-Hazmi MA: Genetic diseases in Arab populations. East Mediterr Health J 2000;6:1102–1103.
10 Zahed L: The spectrum of beta-thalassemia mutations in the Arab populations. J Biomed Biothechnol 2001;1:129–132.
11 Majeed HA, El-Shanti H, Al-Khateeb MS, Rabaiha ZA: Genotype/phenotype correlations in Arab patients with familial Mediterranean fever. Semin Arthritis Rheum 2002;31: 371–376.
12 Gershoni-Baruch R, Shinawi M, Leah K, Badarnah K, Brik R: Familial Mediterranean fever: Penetrance and genetic draft. Eur J Hum Genet 2001;9:634–637.
13 Salih MA, Mahdi AH, Al-Jarallah AA, Al-Jarallah AS, Al-Saadi H, Hafeez MA, Aziz SA: Childhood neuromuscular disorders: A decade's experience in Saudi Arabia. Ann Trop Paediatr 1996;16:271–280.
14 Al-Gazali LI, Bakir M, Hamid Z, Varady E, Varghes M, Haas D, Bener A, Padmanabham R, Abdulrrazzq YM, Dawadu A, Abdulrrazzaq YM, Dawodu AK: Birth prevalence and pattern of osteochondrodysplasias in an inbred high risk population. Birth Defects Res Part A Clin Mol Teratol 2003;67:125–132.
15 Dagoneau N, Scheffer D, Huber C, Al-Gazali LI, Di Rocco M, et al: Null leukemia inhibitory factor (LIFR) mutations in Stüve-Wiedemann/Schwartz-Jampel type 2 syndrome. Am J Hum Genet 2004;74:298–305.
16 Abdul Wahab A, Al-Thani G, Dawod ST, Kambouris M, Al-Hamed M: Heterogeneity of the cystic fibrosis phenotype in a large kindred family in Qatar with cystic fibrosis mutation (I1234V). J Trop Pediatr 2001;47:110–112.
17 Laufer-Cahana A, Lerer I, Sagi M, Rachmilewitz-Minei T, Zami C, Rivlin JR, Abeliovich D: Cystic fibrosis mutations in Israeli Arab patients. Hum Genet 1999, 14:543.
18 Kambouris M, Banjar H, Mogarri I, Nazer H, Al-Hamed M, Meyer BF: Identification of novel mutations in Arabs with cystic fibrosis and their impact on cystic fibrosis transmembrane regular mutation rate in Arab populations. Eur J Pediatr 2000;159:303–309.
19 Messaoud T, Verlingue C, Denamur E, Pascand O, Quere I, Fattoum S, Elion J, Ferec C: Distribution of CFTR mutations in cystic fibrosis patients of Tunisian origin: Identifications of two novel mutations. Eur J Hum Genet 1996;4:20–24.
20 Sheffield VC, Nishimura D, Stone EM: The molecular genetics of Bardet-Biedl syndrome. Curr Opin Genet Der 2001;11:317–321.
21 Shahin H, Walsh T, Sobe T, Lynch E, King MC, Avraham KB, Kanaan M: Genetics of congenital deafness in the Palestinian populations: Multiple connexin 26 alleles with shared origins in the Middle East. Hum Genet 2002; 110:284–289.
22 Abdul Wahab A, Janahi IA, Eltohami A, Zeid A, Ul-Haque NF, Teebi AS: A new type of Ehlers-Danlos syndrome associated with tortuous systemic arteries in a large kindred from Qatar. Acta Paediatr 2003;92:456–462.

Community Genetics

Community Genet 2005;8:27–30
DOI: 10.1159/000083334

Common Autosomal Recessive Diseases in Oman Derived from a Hospital-Based Registry

A. Rajab[a] B. Bappal[b] H. Al-Shaikh[b] S. Al-Khusaibi[b] A.J. Mohammed[a]

[a]Genetic Unit, Directorate General of Health Affairs, Ministry of Health, [b] Pediatric Department, Royal Hospital, Muscat, Sultanate of Oman

Key Words

Autosomal recessive disorders · Birth prevalence · Sultanate of Oman

Abstract

Objectives: The local frequencies of genetic disorders in Oman apart from hemoglobinopathies are largely unknown. The aim of the present study was to evaluate birth prevalence of commonly diagnosed autosomal recessive diseases and to estimate needs and priorities of genetic services. ***Methods:*** Analysis of the years 1993–2002 using a hospital-based registry of genetic diseases was attempted. More than 3,000 records were reviewed. Only patients with definite diagnosis were included in the analysis. Genetically determined diseases occurring less frequently than 1 in 50,000 births are not included. ***Results:*** A number of rare autosomal recessive diseases are found to have a prevalence at least 1 in 50,000 livebirths. ***Conclusions:*** The data suggest that genetic diseases are important as major contributors to perinatal and childhood mortality and morbidity. The need for preventive genetic service is essential for the health of the community in Oman. Autosomal recessive diseases were frequently concentrated in specific geographical areas, which can be explained by founder effect and genetic drift. However, the hospital-based registry may present incomplete information. Further prospective studies are needed to provide more detailed data.

Introduction

The revolution in DNA technology has completely altered the understanding of genetic diseases and the practice of the medical profession.

Rare or 'orphan' diseases are defined as those affecting fewer than 200,000 people in the United States (www.rarediseases.org). Over 6,000 rare diseases have been documented, with the genetic basis for many now well understood. There are numerous observations that autosomal recessive conditions are common in Middle Eastern states [1–4]. The causes for a higher frequency of autosomal recessive diseases are believed to be the high level of consanguinity and the presence of hemoglobin disorders [1]. We present birth prevalence of commonly diagnosed autosomal recessive diseases in the Sultanate of Oman derived from a hospital-based register for1993–2002.

KARGER

Fax + 41 61 306 12 34
E-Mail karger@karger.ch
www.karger.com

1422–2795/05/0081–0027$22.00/0

Accessible online at:
www.karger.com/cmg

Dr. A. Rajab
Genetic Unit, Directorate General of Health Affairs
Ministry of Health
Muscat (Sultanate of Oman)
E-Mail drarajab@omantel.net.om

Table 1. Commonly ascertained autosomal recessive diseases in the Sultanate of Oman among 420,000 livebirths (1993–2002)

Autosomal recessive diseases	Patients born 1993-2002	Observed incidence
Spinal muscular atrophy (Wernig-Hoffman disease)	56	1 in 10,000 births
Congenital adrenal hyperplasia	55	1 in 10,000 births
Polycystic kidney disease	34	1 in 12,000 births
Cystic fibrosis	32	1 in 15,000 births
Primary microcephaly	31	1 in 15,000 births
Renal tubular acidosis	28	1 in 20,000 births
Congenital nephrotic syndrome (Finish type)	25	1 in 20,000 births
Nesidioblastosis	24	1 in 20,000 births
Apple peel bowel syndrome	21	1 in 20,000 births
Zellweger syndrome	19	1 in 20,000 births
Metachromatic leukodystrophy	18	1 in 25,000 births
Congenital generalized lipodystrophy	18	1 in 25,000 births
Ellis-Van Creveld syndrome	18	1 in 25,000 births
Schwartz-Jampel syndrome	15	1 in 30,000 births
Bardet-Biedl syndrome	14	1 in 30,000 births
Robinow syndrome	12	1 in 35,000 births
Oculocutaneous albinism	14	1 in 30,000 births
Epidermolysis bullosa	15	1 in 30,000 births
Galactosialidosis	9	1 in 50,000 births
Cerebro-oculo-musculo-skeletal syndrome	9	1 in 50,000 births
Meckel-Gruber syndrome	9	1 in 50,000 births
Carbohydrate-deficient glycoprotein syndrome	8	1 in 50,000 births
Mucopolysaccharidosis	8	1 in 50,000 births

Subjects and Methods

The Sultanate of Oman has a population of 2 million and a comprehensive healthcare system with specialized genetic service being based in the tertiary referral hospital in the capital Muscat. From 1993 to 2002, clinical data for children attending clinical genetic clinics were prospectively collected. The information included clinical examination, full pedigrees, tribal origin, birth history, growth charts, developmental assessment, radiological and laboratory investigations including histological and enzymatic essays. It was not possible to obtain full investigation in all cases because of the geographic distances, failure to attend for follow-up and limited access to some investigations. The computerized data on autosomal recessive conditions for 1993–2002 were sorted by diagnosis and cross-checked with patient folders. Patients with unresolved diagnosis were excluded from the records. The number of individual diseases seen in children born from 1993 to 2002 was compared to the number of livebirths in the same period. Only conditions with a prevalence of 1 in 50,000 livebirths or higher were recorded. Information on commonly seen autosomal recessive diseases was analyzed with respect to the place of residence in order to look for the geographic clusters.

Results

Most frequently recorded autosomal recessive diseases in the Sultanate of Oman from the hospital-based data of 1993–2002 are presented in table 1.

Geographical areas with high birth prevalence of specific autosomal recessive disorders are depicted in figure 1.

Discussion

The incidence of individual recessive conditions assessed on the basis of a hospital-based register is unusually high in some geographical areas of Oman (table 1). The true figures may be much higher as not all children with genetic diseases may have been referred to genetic clinics. The postmortem examination in stillbirths and children dying in the first few years of life has not been given importance yet, which results in an underestimation of genetic disease. Large groups of recessively inherited neurological, dermatological, ophthalmologic, surgical, orthopedic and many other disorders are not included in this analysis and would need further study.

In Oman, consanguineous unions are common [5], and genetic drift has led to a concentration of genes for specific disorders in tribes and certain population groups. There are a number of communities in the Middle East and around the world with similar observations [6–10].

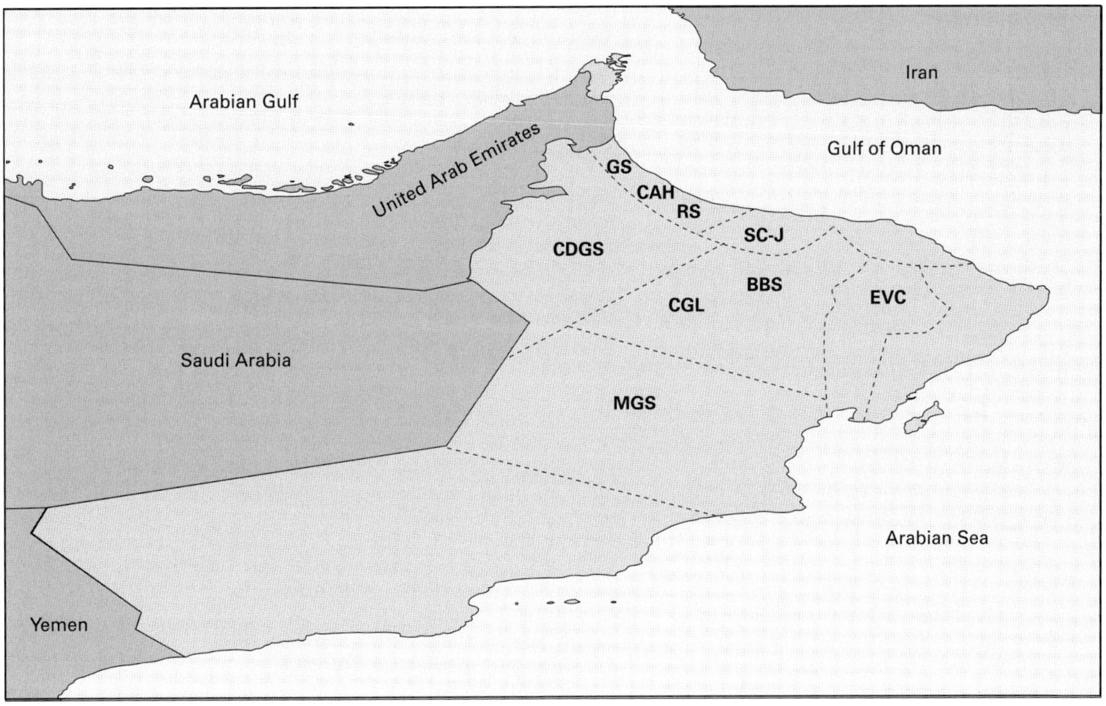

Fig. 1. Geographical areas of Oman representing high density of (90% of cases) autosomal recessive conditions. The northern parts of Oman are densely populated areas. The southwestern parts of Oman are nomadic desert areas with a sparse population. BBS = Bardet-Biedl syndrome; CAH = congenital adrenal hyperplasia; CDGS = carbohydrate-deficient glycoprotein syndrome; CGL = congenital generalized lipodystrophy; EVC = Ellis Van Creveld syndrome; GS = galactosialidosis; MGS = Meckel-Gruber syndrome; RS = Robinow syndrome; SC-J = Schwartz-Jampel syndrome.

Since most of the disease genes originated from the small original founding group, it could be a possibility that most of the carriers in a tribal population have exactly the same mutation at the DNA level. If this situation will prove to be true it will offer an opportunity for rationalizing control of these disorders, and plans to provide population screening for common recessive disorders using modern DNA microchip methods. However, available data suggest that it may not be true for all disorders. The recent studies in Old Order Mennonites suggested that allelic diversity may not be lost in genetic isolates and inbreeding only affects genotype frequencies [9].

Only a few autosomal recessive diseases from Oman were studied on a molecular level: in some a single mutation [11–15] and in others a wide range of mutations were found [16–18].

The availability of information may affect health care decisions. In the absence of central notification systems and using limited medical records, the accrual of epidemiological data is a great challenge. The reliability of data on local frequencies of genetic disorders largely depends on local clinical expertise and – to some extent – on the supporting laboratory facilities. It becomes a vicious circle in such a situation, as it is not possible to diagnose many genetic diseases without adequate facilities and it seems that genetic diseases do not make a significant impact on health because they are not diagnosed. Therefore, the development of national expertise is important.

The genetic structure of populations plays a very significant role in the distribution patterns of specific inherited disorders, and there is a marked difference in the pattern of inherited diseases between developing and developed countries [19]. An important challenge for the years to come is to construct an efficient program for the prevention of genetic diseases.

References

1 Alwan A, Modell B: Community control of genetic and congenital disorders. EMRO Technical Publications Series 24. Alexandria, WHO Eastern Mediterranean Regional Office, 1997.

2 Teebi AS, Farag TI: Genetic disorders in Arab populations. Oxford Monographs on Medical Genetics No. 30. Oxford, Oxford University Press, 1997.

3 Teebi AS: Autosomal recessive disorders among Arabs: An overview from Kuwait. J Med Genet 1994;31:224–233.

4 Al Hosani H, Czeizel AE: Unique demographic situation in the United Arab Emirates. Am J Med Genet 1996;61:1.

5 Rajab A, Patton M: A study of consanguinity in the Sultanate of Oman. Ann Hum Biol 2000;3: 321–326.

6 Al-Odaib AN, Abu-Amero KK, Ozand PT, Al-Hellani AM: A new era for preventive genetic programs in the Arabian Peninsula. Saudi Med J 2003;24:1168–1175.

7 Bundey S, Alam H: A five-year prospective study of the health of children in different ethnic groups, with particular reference to the effect of inbreeding. Eur J Hum Genet 1992; 24:1–15.

8 Peltonen L, Pekkarinen P, Aaltonen J: Messages for an isolate: Lessons from the Finish gene pool. Biol Chem Hoppe Seyler 1995;376: 697–707.

9 Puffenberger EG: Genetic heritage of the Old Order Mennonites of southeastern Pennsylvania. Am J Med Genet Part C 2003;121C:18–31.

10 Verma IC, Bijarnia S: The burden of genetic disorders in India and a framework for community control. Community Genet 2002;5: 192–196.

11 Afzal AR, Rajab A, Fenzke CD, Oldridge M, Elanko N, Ternes-Pereira E, Tuysuz B, Murday V, Patton MA, Wilkie A, Jeffery S: Recessive Robinow syndrome, allelic to dominant brachydactily type B, is caused by mutation of ROR2 (letter). Nat Genet 2000;25:419–422.

12 Afzal AR, Rajab A, Fenske C, Crosby A, Lahiri N, Ternes-Pereira E, Murday VA, Houlston R, Patton MA, Jeffrey S: Linkage of recessive Robinow syndrome to a 4 cM interval on chromosome 9q22. Hum Genet 2000;106:351–354.

13 Nicole S, Davoine CS, Topaloglu H, Cattolico L, Barral D, Beighton P, Ben Hamida C, Hammouda H, Cruaud C, White PS, Samson D, Urtizberea JA, Lehmann-Horn F, Weissenbach J, Hentati F, Fontaine B: Perlecan, the major proteoglycan of basement membranes, is altered in patients with Schwartz-Jampel syndrome (chondrodystrophic myotonia). Nat Genet 2000;26:480–483.

14 Rajab A, Kunze J, Mundlos S: Spondyloepiphyseal dysplasia Omani type. A new recessive type of SED with progressive spinal involvement. Am J Med Genet 2004;126A:413–419.

15 Tiele H, Sakano M, Kitagawa K, Rajab A, Hohne W, Ritter H, Leschik G, Nurnberg P, Mundlos S: Loss of chonroitin 6-O-sulfotransferase-1 function results in severe human chondrodysplasia with progressive spinal involvement. Proc Natl Acad Sci USA 2004;101: 10155–10160.

16 Frossard PM, Girodon E, Dawson KP, Ghanem N, Plassa F, Lestringant GG, Gooseens M: Identification of cystic fibrosis mutations in the United Arab Emirates. Mutations in brief No 133. Online. Hum Mutat 1998;11:412–413.

17 Rajab AG, Patton MA, Modell B: Study of hemoglobinopathies in Oman through a national register. Saudi Med J 2000;21:1168–1172.

18 Rajab A, Heathcote K, Joshi S, Jeffery S, Patton M: 2002 Heterogeneity for congenital generalized lipodystrophy in seventeen patients from Oman. Am J Med Genet 2002;110:219–225.

19 Bittles AH, Savithri HS, Rao NA: Community genetics in developing countries (editorial). Community Genet 2002;5:151–152.

Community Genetics

Community Genet 2005;8:31–34
DOI: 10.1159/000083335

Genetic Contribution to High Neonatally Lethal Malformation Rate in the United Arab Emirates

A. Dawodu[a] L. Al-Gazali[a] E. Varady[b] M. Varghese[c] K. Nath[d] V. Rajan[d]

Departments of Pediatrics, [a]Faculty of Medicine and Health Sciences, United Arab Emirates University, [b]Tawam Hospital, [c]Oasis Hospital, and [d]Al Ain Hospital, Al Ain, United Arab Emirates

Key Words
Genetics · Lethal malformations · Neonates · United Arab Emirates

Abstract
Objectives: We examined the contribution of genetic disorders to congenital anomalies (CA) causing neonatal deaths in the Al Ain Medical District (AMD) in the United Arab Emirates (UAE) because of the high consanguineous marriage rate in the community. ***Methods:*** Charts of all neonatal deaths in the three perinatal units, which accounted for 99% of all births in AMD (1992–2000), were studied. Data regarding pregnancy, a family history including the level of parental consanguinity, the results of genetic evaluations and neonatal outcomes were recorded as part of an ongoing malformation surveillance system. Causes of death were based on clinical, laboratory and imaging findings. ***Results:*** Of the 508 neonates who died, 212 (42%) had CA, which were the leading cause of death. Forty-four percent of the CA were due to definite genetic disorders and 75% of these were single gene defects. Multisystem malformations were the commonest congenital malformations. Parental consanguinity was associated with a 2-fold increased risk of nonchromosomal multisystem malformations. ***Conclusions:*** Lethal malformations were the leading cause of neonatal deaths, and parental consanguinity was associated with an increased risk of autosomal recessive disorders. The results underscore the importance of genetic screening and counseling in strategies for further significant reductions in the neonatal mortality rate in the UAE.

Introduction

Recent reports indicated an increasing contribution of congenital anomalies (CA) to infant mortality worldwide and a need for identification of the risk factors and prevention of CA [1]. In the United Arab Emirates (UAE), CA are the leading cause of infant m2005;8:31–34ortality, and 70% of the deaths occur in the neonatal period [2]. In spite of this high neonatal death rate from CA, there have been no reports, to our knowledge, from the UAE on the risk factors for neonatally lethal CA. In a previous study, CA were the second most common cause of neonatal deaths in the Al Ain Medical District (AMD) in the UAE [3]. A majority of the lethal anomalies were multiple or complex and were due to autosomal recessive disorders [3], which differs from western countries [4, 5]. In view of the high rate of consanguineous marriages in the com-

KARGER
Fax +41 61 306 12 34
E-Mail karger@karger.ch
www.karger.com

1422–2795/05/0081–0031$22.00/0

Accessible online at:
www.karger.com/cmg

Adekunle Dawodu, FRCP (E), FRCP (L), FRCPCH, International Health Program
Center for Epidemiology and Biostatistics MLC 5041
Cincinnati Children's Hospital Medical Center
3333 Burnet Avenue, Cincinnati, OH 45229 (USA)
Tel. +1 513 636 1966, E-Mail adekunle.dawodu@cchmc.org

Table 1. Components of neonatally lethal anomalies

Anomalies	n	% of total	n/1,000 LB
Single system	90	42.5	1.3
CNS	37	17.5	0.53
Cardiovascular	16	7.5	0.23
Pulmonary	17	8.0	0.24
Others	20	9.5	0.29
Multiple	122	57.5	1.8
Recognized syndromes	38	17.9	0.55
Non-recognized syndromes	18	8.5	0.26
Chromosomal	24	11.3	0.35
Sequences	26	12.3	0.38
Unidentified	12	5.7	0.17
Associations	4	1.9	0.06
Total	212	100	3.1

munity [6], parental consanguinity was suggested as an important contributing factor to the high rate of lethal autosomal recessive disorders [3].

As part of an ongoing neonatal audit, we examined the pattern of CA causing neonatal deaths. We also quantified the contributions of definite genetic disorders and parental consanguinity to neonatally lethal anomalies in the district.

Materials and Methods

All neonatal deaths (birth weight ≥500 g and age 0–27 days) in AMD in the UAE were studied during the period 1992–2000. The three contributory perinatal centers accounted for 99% of the births in AMD. AMD has a population of 400,000 and between 8,000 and 8,500 annual births [2]. The cause of death in each case was reviewed by at least one of the authors and classified according to accepted criteria [3]. As in a previous report, the cause of death was based on clinical, laboratory and imaging findings [3]. One of the authors (L.A.G.), a clinical geneticist, was involved in the evaluation of CA cases.

For CA cases, detailed pregnancy and family histories, including the levels of parental consanguinity, were recorded as part of an ongoing malformation surveillance system in AMD. Definite genetic contributions were examined and classified as chromosomal anomalies and single gene disorders. The latter were classified in the Mendelian fashion. Statistical analyses were carried out when appropriate, and a p value <0.05 was considered significant.

Results

There were 69,209 live births (LB) and 508 neonatal deaths, resulting in a neonatal mortality rate of 7.3 per 1,000 LB during the study period. Of the 508 neonatal deaths, 212 (42%) were lethal anomalies, while problems of preterm births, bacterial sepsis/necrotizing enterocolitis, asphyxia and other conditions accounted for 27, 14, 10 and 7% of the neonatal deaths, respectively. The birth prevalence of neonatally lethal malformation was 31 per 10,000 LB.

Type of Anomalies

A review of the pattern of anomalies (table 1) showed that 122 (58%) were multiple CA and 90 (42%) were single system malformations. Central nervous system anomalies were the most common single system anomalies and included anencephaly (14), hydrocephalus (8), spinal muscular atrophy (4), neural tube defects (3) and others (8). Eleven of the cardiac malformations were classified as complex malformations by echocardiography. Thirteen of the 17 pulmonary malformations were secondary to diaphragmatic hernia, or congenital hypoplastic lungs. The other single system anomalies included gastrointestinal malformations (5), isolated renal malformations (4), epidermolysis bullosa (3), anomalies secondary to intrauterine infections (4) and unclassified cases (4).

Of the 122 multiple anomalies, 56 (45%) were nonchromosomal multisystem malformations of which the majority (67%) were recognized syndromes due to single gene defects. The chromosomal anomalies included trisomy 18 (12), trisomy 13 (4) and trisomy 21 (5), translocations and triploidy. The cases of sequences in which the anomalies could be regarded as secondary to a primary malformation were Potter's (20) and prune belly (4) syndromes, and holoprosencephaly (2). There were 12 cases in which the pattern was unspecified.

Genetic Contributions

Definite genetic disorders accounted for 94 (44%) of the lethal anomalies. 24 (11%) were associated with chromosomal anomalies and 70 (33%) were due to single gene defects. A majority (80%) of the single gene defects were autosomal recessive disorders, many of which were very rare or not previously reported. There were 11 cases of autosomal dominant disorders, 8 of which were new mutations. The incidence of neonatally lethal autosomal recessive disorders was 8.5 per 10,000 LB, while the incidence of lethal chromosomal anomalies was 3.5 per 10,000 LB.

Table 2. Malformations and parental consanguinity

	Consanguineous n = 101	Non-consanguineous n = 42
Single		
CNS	18	8
Cardiac	2	2
Pulmonary	4	6
Renal	2	0
Others	6	2
Multiple		
Multisystem malformations	44 1.98 (1.07–3.68)[a]	9
Chromosomal	11	7
Sequences	11	6
Associations	2	1
Unidentified	1	1

[a] RR (95% CI); p = 0.016.

The possible contribution of parental consanguinity to the high incidence of autosomal recessive disorders was assessed by comparing the pattern of lethal anomalies by parental consanguinity status. Information on parental consanguinity was available in 143 of the 212 cases of lethal malformations. Of these, 101 (70%) were consanguineous, of which 60% were first cousin marriages. There was a two-fold increased risk of non-chromosomal lethal multi-system malformations in the offspring of consanguineous parents (relative risk: 1.98, 95% confidence interval 1.07–3.68, $p = 0.016$; table 2). Most of these non-chromosomal multisystem malformations in the offspring of consanguineous parents were autosomal recessive disorders.

Discussion

The neonatal mortality rate in this study is similar to the figures from developed countries but lower than those from other developing countries [7]. Similar to studies from western countries [5], lethal malformations are currently the leading cause of neonatal deaths; however, the birth prevalence of neonatally lethal malformations is higher than that reported from these countries. For example, the lethal malformation rate of 31 per 10,000 LB in this study is significantly higher ($p < 0.0001$) than the rate of 15 per 10,000 LB in a similar study of 337 neonatal deaths in Iowa, USA [8]. This finding suggests that a significant reduction in neonatal and infant mortality rates in the UAE would require the prevention of lethal CA. This is in agreement with recent reports from the USA [9] and an international assessment of the contribution of CA to infant mortality [1], which indicated a need to focus on a reduction in congenital malformations, since they constitute the leading cause of infant deaths [9].

The difference between the pattern of neonatally lethal malformations in this study and similar studies from North America was the higher birth prevalence of multiple anomalies, mostly due to excess lethal autosomal recessive disorders. For example, 57% of the lethal malformations in this study were multisystem anomalies, while similar analyses of causes of neonatal deaths from the USA found only 10 [4] and 29% [10] of lethal multiple malformations. Furthermore, 38 (68%) of 56 non-chromosomal multisystem malformations were due to recognized syndromes, while recognized syndromic malformations were very rarely seen in a similar study from neonatal intensive care in the USA [10]. The pattern of lethal malformation in this study is, however, similar to the general CA profile previously reported for this community [11]. Studies of perinatal deaths in the UK also noted a similar pattern of a higher birth prevalence of lethal multisystem anomalies among South Asians, especially Pakistanis, compared with Europeans [12–14].

Genetic factors have been implicated as contributing to ethnic differences in the pattern of lethal malformations, but there have been few studies with detailed analysis of the specific genetic disorders [12–14]. Furthermore, most of the previous studies on genetic factors examined perinatal deaths [12–15], and in the only previous study of neonatal deaths the rate of genetic evaluation was low [10]. In contrast, a clinical geneticist was involved in the evaluation of almost all the cases in the present investigation, which may have contributed to a more accurate assessment of genetic disorders. About half of the lethal malformations observed in the present study were associated with definite genetic disorders, of which 75% were due to single gene defects, mostly autosomal. A similar higher birth prevalence of lethal autosomal recessive disorders was reported in perinatal deaths in Pakistanis when compared with Europeans or Indians [12–14] and attributed to the high parental consanguinity rate among the Pakistanis [12–15].

Where information was available, 70% of the parents of infants with lethal malformations were consanguineous and 60% of these were first cousin marriages, compared with 50% consanguineous marriage and 26% first cousin marriage rates in the general population [6]. Furthermore, the prevalence of non-chromosomal multisystem malformations, both syndromic and non-syndromic, was significantly higher in the offspring of consanguineous parents compared with non-consanguineous parents. The study confirms that parental consanguinity accounted in part for the excess neonatally lethal malformations in this community.

The birth prevalence of neonatally lethal chromosomal disorders in this study was similar to those reported in studies of perinatal deaths in Europeans and South Asians resident in the UK, which suggested no ethnic differences in the rates of lethal chromosomal anomalies [12, 13].

In conclusion, CA, mostly multiple, were the leading cause of neonatal deaths in this UAE study. The birth prevalence of lethal autosomal disorders was high and parental consanguinity was associated with increased risks of syndromic and non-syndromic multisystem anomalies. Therefore, accurate diagnosis and effective genetic screening and counseling are essential for further significant reductions in neonatal mortality. The findings of this study may have applications in other Arab countries with similar cultural risk factors.

References

1 Rosana A, Botto LD, Botting B, Mastroiacovo P: Infant mortality and congenital anomalies from 1950 to 1994: An international perspective. J Epidemiol Community Health 2000;54:660–666.

2 Ministry of Health: Annual Report: Department of Preventive Medicine. Abu Dhabi, Ministry of Health, 2002.

3 Dawodu A, Varady E, Verghese M, Al-Gazali LI: Neonatal audit in the United Arab Emirates: A country with a rapidly developing economy. East Mediterr Health J 2000;6:55–64.

4 Phillip AGS: Neonatal mortality rate: Is further improvement possible? J Pediatr 1995;126:427–433.

5 Foran A, Dempsey E, Watters A, Gormally SM: Irish neonatal mortality – 12 years on. Ir Med J 2002;95:267–270.

6 Al-Gazali LI, Bener A, Abdulrazzaq YM, Micallef R, Al-Khayat AI, Gaber T: Consanguineous marriages in the United Arab Emirates. J Biosoc Sci 1997;29:491–497.

7 Yu VYH: Global, regional and national perinatal and neonatal mortality. J Perinat Med 2003;31:376–379.

8 Hein HA, Lofgren M: The changing pattern of neonatal mortality in a regionalized system of perinatal care: A current up-date. Pediatrics 1999;104:1064–1069.

9 Malcoe LH, Shaw GM, Lammer EJ, Herman AA: The effect of congenital anomalies on mortality risk in white and black infants. Am J Public Health 1999;89:887–892.

10 Hudome SM, Kirby RS, Senner JW, Cunniff C: Contribution of genetic disorders to neonatal mortality in a regional intensive care setting. Am J Perinatol 1994;11:100–103.

11 Al-Gazali LI, Dawodu AH, Sabrinathan M, Varghese M: The profile of major congenital abnormalities in the UAE population. J Med Genet 1995;32:7–13.

12 Young ID, Clarke M: Lethal malformations and perinatal mortality: A 10-year review with comparison of ethnic differences. Br Med J 1987;295:89–91.

13 Chitty LS, Winter RM: Perinatal mortality in different ethnic groups. Arch Dis Child 1989;64:1036–1041.

14 Terry PB, Bissenden JG, Condie RG, Mathew PM: Ethnic differences in congenital malformations. Arch Dis Child 1985;60:866–879.

15 Bundey S, Alam H, Kaur A, Mir S, Lancashire R: Why do UK-born Pakistani babies have high perinatal and neonatal mortality rates? Paediatr Perinat Epidemiol 1991;5:101–104.

Community Genetics

Community Genet 2005;8:35–39
DOI: 10.1159/000083336

Molecular Heterogeneity of Beta-Thalassemia in the United Arab Emirates

E. Baysal

Dubai Genetic and Thalassemia Center, Al Wasl Hospital, Dubai, United Arab Emirates

Key Words
Consanguinity · β-Thalassemia · United Arab Emirates

Abstract
The β-thalassemia alleles in 313 national patients of the United Arab Emirates (UAE) have been characterized using PCR-based DNA-diagnostic techniques including DNA sequencing. A total of 212 patients had homozygous β-thalassemia and the remaining 101 were compound heterozygotes. More than half of the patients were homozygous for the IVS-I-5 (G→C) mutation followed by the sickle cell gene. The latter accounted for 25% of the chromosomes. In terms of frequency, five β-thalassemia mutations; IVS-I-5 (G→C), βS, –25 bp del, Cd 8/9 (+G) and IVS-II-1 (G→A) accounted for 83% of the alleles. In addition, 427 expatriate patients were studied: 256 with homozygous β-thalassemia and 171 were compound heterozygotes. In both the UAE nationals and expatriates, the β-thalassemia mutations and their frequency followed a similar trend. Our results indicate that the frequency of β-globin gene defects including β-thalassemia, sickle cell gene (βS) and abnormal hemoglobins is significantly increased and poses a major public health problem in the UAE. The number of homozygous patients strongly suggests a high degree of consanguinity among the UAE nationals. With 50 different β-thalassemia alleles, UAE is arguably the most heterogeneous population in the world. The diversity of these mutations reflects the historical admixture of genes and their migration from different areas in the region. Our data strongly suggest the need for a comprehensive thalassemia control program and provides a basis for population screening, genetic counseling and prenatal diagnosis.

Introduction

The population of the United Arab Emirates (UAE) is slightly over 4 million but only 20% are indigenous nationals. The birth rate is estimated at 2%, with infant mortality rate <1%. The fertility rate is 4.04 births per national woman. A noteworthy feature of the UAE population is its age structure; the 25–29 age group is the largest (500,000), followed by 30–34 age group (460,000), and 35–39 age group (430,000). Those in the 10–19 age group (510,000) comprise 15% of the entire population. Those under the age of 50 account for 95% of the total population. In the 20–49 age group, males account for nearly three times the number of females. Approximately 80% of the UAE population are made up of the expatriate community, the majority being Indians, Pakistanis, Iranians

KARGER

Fax +41 61 306 12 34
E-Mail karger@karger.ch
www.karger.com

1422–2795/05/0081–0035$22.00/0

Accessible online at:
www.karger.com/cmg

Dr. Erol Baysal
Consultant Molecular Biologist, Head, Molecular Genetic Unit
Dubai Genetic and Thalassemia Center
Al Wasl Hospital, Dubai (United Arab Emirates)
Tel. +971 4306 3244, Fax +971 4324 2511, E-Mail emerinah@emirates.net.ae

and other Arab nationals. The term 'expatriate' will be used collectively to represent the above ethnic groups.

Autosomal recessive disorders are common in the UAE. The most common disorders include the hemoglobinopathies, metabolic diseases and neurodegenerative disorders. Monogenic diseases such as cystic fibrosis, fragile-X syndrome and Tay-Sach disease are also believed to exist at appreciable levels. During the last year alone, the author's laboratory has identified and characterized more than 50 cases of cystic fibrosis, predominantly in UAE nationals.

β-Thalassemia constitutes a major public health problem in the UAE. From 1989 to 2003, more than 850 patients have been registered at the Dubai Genetic and Thalassemia Center. The exact number of thalassemics in the UAE is unknown but DNA-based data from the author's laboratory project this number to be much higher when the other emirates are taken into account. Previous surveys showed that the UAE exhibit one of the highest carrier frequencies of β-thalassemia in the Gulf region. Studies performed on over 400 UAE national newborns, nearly 2,000 adult college students and 800 randomly selected nationals demonstrated that the incidence of β-thalassemia was approximately 8.5% [1]. In the UAE, sickle cell gene (βS) also contributes significantly to the genetic epidemiology of hemoglobinopathies. It is a major genetic factor in a group of patients with co-inherited Sβ-thalassemia genotype. Our DNA-based studies demonstrated that the βS gene is indeed the second most common β-globin gene defect in the UAE [2]. In addition HbE is also present particularly in the expatriate community from Bangladesh. It is important to note that the majority of the β-thalassemia mutations in the UAE are very severe. Except for the most common allele (which is β^+ thalassemia), all other mutations are of severe β^0-thalassemia type. The high degree of consanguinity, especially between the first cousin marriages, resulted in a significant number of homozygotes who are on regular blood transfusion and iron chelation therapy. Mutation analysis among the UAE nationals and expatriate β-thalassemia patients clearly demonstrated that the UAE is arguably the most heterogenous β-thalassemia population in the world with 50 different β-thalassemia mutations reported to date. Among the UAE nationals, a total of 212 patients had homozygous β-thalassemia. This was more than twice the number of compound heterozygotes. Public awareness programs, education and preventive measures have recently been implemented and seem to be showing some positive effect. The findings presented here indicate the need for a comprehensive thalassemia control program and provide a foundation for patient management, population screening, genetic counseling and prenatal diagnosis.

Molecular Studies

These are carried out routinely on all β-thalassemia patients who are registered at the Genetic and Thalassemia Center. α-Globin genes were routinely investigated as the latter exist at a very high frequency [3]. The molecular analyses include isoelectric focusing, quantitation of Hb types by column chromatography, PCR, restriction enzyme analysis, β-strip hybridization, allele-specific oligonucleotide hybridization as well as manual and automated DNA sequencing.

Patients' DNA were extracted according to the previously described methods [4]. The 5′β segment of the β-globin gene was amplified using a forward primer, located in the upstream promoter region 5′ to the Cap site, and a reverse primer in the second intervening sequence. A vast majority of the β-thalassemia mutations that exist in the UAE are found in the 5′β segment. The 3′β segment was only amplified whenever mutation screening in the 5′β revealed no mutation.

Results

UAE Nationals

A total of 313 (212 homozygous and 101 compound heterozygous) patients were studied. The number of homozygous β-thalassemia patients was more than double the compound heterozygotes. The β-thalassemia spectrum in the UAE nationals showed considerable homozygous mutations strongly suggesting parental consanguinity. Table 1 shows the distribution of homozygous β-thalassemia mutations among UAE national patients up to January 2004. The IVS-I-5 (G→C) homozygosity surpasses all other mutations by a large margin. In fact, more than half (54.7%) of our patients were diagnosed as homozygous for the IVS-I-5 (G→C) mutation. This is perhaps one of the highest prevalences in the IVS-I-5 (G→C) allele reported thus far in any world population. This was followed by sickle cell anemia; 25% of our patients were diagnosed with homozygous βC chromosome. Both the IVS-I-5 (G→C) and the βS gene accounted for 80% of the β-globin gene defects among the homozygous β-thalassemia patients from the UAE. Five homozygous mutations, IVS-I-5 (G→C), βS, –25 bp del, Cd 8/9 (+G), and IVS-II-1 (G→A) accounted for 94% of the 212 UAE national β-thalassemia patients. The remaining are the

Table 1. Homozygous β-thalassemia mutations in the UAE (January 2004)

No.	Mutation	n	%
1	IVS-I-5 (G→C)/IVS-I-5 (G→C)	116	54.7
2	Hb S/Hb S	52	24.9
3	Cd8/9 (+G)/Cd8/9 (+G)	11	5.2
4	–25 bp del/–25 bp del	9	4.2
5	Cd39 (C→T)/Cd39 (C→T)	6	2.8
6	IVS-II-1 (G→A)/IVS-II-1 (G→A)	4	1.9
7	Cd30 (G→C)/Cd30 (G→C)	3	1.4
8	Cd5 (–CT)/Cd5 (–CT)	3	1.4
9	Hb D/Hb D [Cd121 (G→C)]	2	1.0
10	IVS-I-1 (G→A)/IVS-I-1 (G→A)	2	1.0
11	–88 (C→A)/–88 (C→A)	2	1.0
12	Cd8 (–AA)/Cd8 (–AA)	1	0.5
13	Cd82/83 (–G)/Cd82/83 (–G)	1	0.5
Total		212	100

Table 2. Frequency of β-thalassemia alleles in the UAE (626 chromosomes; January 2004)

No.	Mutation	n	%
1	IVS-I-5 (G→C)	298	47.6
2	Hb S	137	21.9
3	–25 bp del	39	6.2
4	Cd8/9 (+G)	27	4.3
5	IVS-II-1 (G→A)	18	2.9
6	Cd8 (–AA)	12	1.9
7	Hb D-Punj ab	10	1.6
8	IVS-I-6 (T→C)	10	1.6
9	Cd5 (–CT)	10	1.6
10	Others	65	10.4
Total		626	100.0

Table 3. Compound heterozygous β-thalassemia mutations among the UAE nationals

SN	Mutation	n
1	Hb S/IVS-I-5	17
2	IVS-I-5/–25 bp	10
3	IVS-I-5/Cd8(–AA)	9
4	IVS-I-5/IVS-I-6	9
5	Hb S/–25 bp del	4
6	–25 bp/IVS-II-1	4
7	Cd30/IVS-I-110	3
8	IVS-I-5/IVS-II-1	3
9	Hb S/Hb D	3
10	Cd8/9/IVS-I-5	3
11	IVS-I-5/Hb D	2
12	–25 bp del/Cd82/83	2
13	Cd5/Cd39	2
14	Hb S/Cd44	2
15	Hb S/Cd82/83	2
16	IVS-I-5/–88 (C→A)	2
17	IVS-I-5/Poly A	2
18	IVS-I-5/Cd110(T→C)	2
19	IVS-II-1 (G→A)/Cd8 (–AA)	1
20	–25 bp del/Cd44	1
21	Cd5/Hb S	1
22	Cd5/Poly A	1
23	Cd39/Hb S	1
24	Cd39/IVS-II-848 (C→A)	1
25	Hb S/Cd37 (G→A)	1
26	Hb S/IVS-II-1	1
27	Hb S/Cd30	1
28	Cd8/9/Cd15	1
29	Cd8/9/Cd36/37 (–T)	1
30	Cd30/–101 (C→T)	1
31	IVS-I-110/Cd5 (–CT)	1
32	IVS-I-5/Cd15	1
33	IVS-I-5/Cd39	1
34	IVS-I-5/Cd44	1
35	IVS-I-5/Cd5	1
36	IVS-I-5/Cd82/83	1
37	IVS-I-6/Hb D	1
38	IVS-II-1/Cd82/83	1
Total		101

–25 bp del, Cd 8/9 (+G), Cd 39 (C→T) and IVS-II-1 (G→A) in decreasing frequency (table 1). To date 13 different homozygous mutations have been defined in the UAE national patient population.

Results in table 2 demonstrate the frequency of β-thalassemia mutations. The IVS-I-5 (G→C) mutation is the most common β-thalassemia mutation among the UAE nationals. Of the 626 chromosomes studied from the UAE nationals, 298 (47.6%) were identified with the IVS-I-5 (G→C) mutation. The βS gene is the second most common allele and accounted for 21.9% followed by the –25 bp del at 6.2%.

In addition, β-thalassemia mutations showed considerable heterogeneity. A total of 38 different compound heterozygosities were observed (table 3). The most prevalent was the IVS-I-5 (G→C)/β^S followed by IVS-I-5

(G→C)/–25 bp del and IVS-I-5 (G→C)/Cd8 (–AA). Some of the mutations were rare and some were observed only once. It is important to note that for both the homozygous and compound heterozygous patients, all the mutations were of β^0-type except for the IVS-I-5 (G→C), which is β^+-thalassemia albeit very severe.

Expatriates

A total of 427 patients, 256 homozygous and 171 compound heterozygous patients, were studied. The expatriates constituted approximately 55% of our patient population. Most of the expatriate patients were from India, Pakistan, Iran, Bangladesh, Gulf Cooperation Countries and other Arab countries. Data presented here represent the collective expatriate population.

The β-thalassemia spectrum among the expatriates was similar to that observed for the UAE nationals. The IVS-I-5 (G→C) homozygosity was the most frequent mutation (49.2%) followed by the β^S gene (15.2%) and Cd8/9 (+G) (9.4%). Among 256 homozygous patients studied as of January 2004, 21 different homozygous mutations were characterized again signifying the effect of consanguineous marriages among the predominantly Muslim patient population. As was the case with the UAE nationals, there was almost double the number of homozygous cases as compared to compound heterozygote patients.

A total of 76 different compound heterozygous mutations were defined (data not shown), making the UAE arguably the most heterogeneous β-thalassemia population in the world.

Discussion

The β-thalassemias are prevalent in the UAE. We had previously reported a high prevalence of β-thalassemia among the UAE nationals and the expatriate population in the UAE [1]. The frequency of the β-globin gene defect including β-thalassemia, β^S gene and abnormal hemoglobins is one of the highest in the Gulf Region.

Our attempts to characterize and identify the β-thalassemia mutations among the β-thalassemia patients at the Dubai Genetic and Thalassemia Center have resulted in some unique observations. A striking feature of this study was the high prevalence of homozygous patients. This is the result of a high degree of consanguinity especially between the first cousins [5, 6]. The high level of endogamy originates from century-old sociocultural and religious traditions in the Arab societies. Similar observations were made in the expatriate patient population most of whom were Muslims.

Our molecular studies on the UAE nationals and expatriate β-thalassemia patients revealed a total of 50 β-globin gene mutations. In contrast to the other countries where only a small spectrum of β-thalassemia alleles occur, UAE appears to have a significant heterogeneity even among the indigenous population. Many of the UAE national families are thought to have their roots in surrounding countries such as Iran and Baluchistan. This is evident from the common mutations and haplotype analysis (data not included).

The most common mutation in the UAE is the IVS-I-5 (G→C), which exists at very high frequencies in the Indian subcontinent and among populations surrounding India. Except for this allele, which is β^+ thalassemia, all the remaining 10 most common mutations were β^0-thalassemia with very severe phenotype.

The spectrum of β-thalassemia mutations in the UAE also represents an extensive admixture of genes from the Mediterranean, Arabian and Indian backgrounds. Among the ten most common mutations, five [Cd39 (C→T), IVS-II-1 (G→A), Cd5 (–CT), IVS-I-1 (G→A) and Cd30 (G→C)] exist at significantly high levels in the Mediterranean countries; three are prevalent in India [IVS-I-5 (G→C), Cd 8/9 (+G) and Hb D], and three occur at appreciable levels in Iran and the Eastern Arabian peninsula [–25 bp del, Cd39 (C→T) and IVS-II-I (G→A)].

It is very likely that the IVS-I-5 (G→C) allele was introduced to the Arabian Peninsula by gene migration from Baluchistan, a region spanning southern Iran, Afghanistan, and Pakistan. It has been established that the IVS-I-5 (G→C) mutation is found mainly in the Chinese, Asian-Indians and peoples of Baluchistan and Pakistan but not among the Middle East Arabs [7]. The propensity of this allele in the Arabian peninsula can be attributed to the population migration from the Indian subcontinent; its low frequency in Kuwait and high frequency in the UAE and Oman favors the speculation that the gene was introduced into the Arabian peninsula across the Straits of Hormuz. This navigational route still constitutes a major trade link between the Indian subcontinent and the Gulf States.

One of the most striking features of the β-thalassemia distribution in the UAE is the diversity of mutations. This is caused by the admixture of genes between different groups in and around the Gulf region, Indian subcontinent, Middle Eastern countries and Africa. In recent times this was exacerbated by the influx of many other nations into the UAE. It is apparent from our studies that gene flow and heterogeneity of β-thalassemia mutations represent complex anthropological influences from the East Mediterranean, Asia, India, sub-Sahara and East Africa corroborating that the diversity of β-thalassemia mutations reflects historical events and gene migration in the region. This report indicates the need for a comprehensive thalassemia control program and provides a basis for patient management, population screening, genetic counseling and prenatal diagnosis.

References

1 Baysal E: Hemoglobinopathies in the United Arab Emirates. Hemoglobin 2001;25:247–253.

2 El-Kalla S, Baysal E: Genotype-phenotype correlation of sickle cell disease in the United Arab Emirates. Pediatr Hematol Oncol 1998; 15:237–242.

3 El-Kalla, Baysal E: α-Thalassemia in the United Arab Emirates. Acta Haematol 1998;100: 49–53.

4 Poncz M, Solowiejczyk D, Harpel B, Mory Y, Schwartz E, Surrey S: Construction of human gene libraries from small amounts of peripheral blood: Analysis of β-like globin genes. Hemoglobin 1982;6:27–36.

5 Bener A, Abdulrazzaq YM, Al-Gazali LI, Micallef R, Al Khayat AI, Gaber T: Consanguinity and associated socio-demographic factors in the United Arab Emirates. Hum Hered 1996;46:256–264.

6 Al-Gazali LI, Bener A, Abdulrazzaq YM, Micallef R, Al Khayat AI, Gaber T: Consanguineous marriages in the United Arab Emirates. J Biosoc Sci 1997,29:491–497.

7 Quaife RQ, Al-Gazali L, Abbas S, Fitzgerald P, Fitches A, Valler D, Old JM: The spectrum of β-thalassemia mutations in the UAE national population. J Med Genet 1994;31:59–61.

Community Genetics

Community Genet 2005;8:40–43
DOI: 10.1159/000083337

DNA Diagnosis of Single Gene Disorders in Patients of Non-European Origin: Experience from Kuwait

E. Samilchuk

Kuwait Medical Genetics Center, Sulaibikhat, Kuwait

Key Words
DNA diagnostics · Genetic disorders · Kuwait

Abstract
Objective: The aim of this study was to establish genetic DNA-diagnostic service in Kuwait. ***Methods:*** Polymerase chain reaction, restriction fragment length polymorphisms, heteroduplex analysis and DNA sequencing were applied. ***Results:*** Direct testing for common mutations had variable success in Kuwaiti patients with different genetic disorders, and additional mutation analysis was required in many cases. Genetic heterogeneity, mutations of Mediterranean, African and Arabic/Middle Eastern origin, and homozygosity by descent are characteristic of patients from Kuwait. ***Conclusions:*** More efforts aimed at the identification of mutations underlying genetic disorders in Kuwait as well as in other Gulf countries are warranted. This can be achieved by focusing genetic research in the academic institutions of Gulf countries towards this goal.

Introduction

Genetic DNA-diagnostic laboratories have become part of the health care system in many countries. The service provided by such laboratories is based on the direct testing for common mutations. The vast majority of disease-causing genes/mutations have been identified in the North American and Western European research centers. As a result, most of the data on pathogenic mutations represent patients of European descent. Since mutation frequencies vary significantly among different populations, data on 'local' variants are essential for establishing similar services outside North America and Western Europe. However, for many non-European populations, these data are lacking.

Kuwait, a small country in the north eastern part of the Arabian Peninsula bordering Iraq and Saudi Arabia, has a very diverse population comprising 1.9 million Kuwaiti nationals, Beduins (stateless Arabs) and expatriates. The Kuwaiti nationals are mostly Arabs, many of whom trace their origin to different Beduin tribes that roamed in central and northern parts of the Arabian Peninsula. However, a significant proportion of Kuwaiti nationals are of Persian descent, and there are also several minor groups, e.g. Armenians and European wives of Kuwaiti nationals. The main expatriate groups in Kuwait are the nationals of other Arab countries (Egypt, Syria and Jordan), as well as Iran, India, Pakistan, Bangladesh, Sri Lanka, the Philippines and Thailand.

Genetic diseases are very common among Arabs living in Kuwait, which has been attributed to the high frequency of consanguineous marriages, usually between first cousins (reviewed by Teebi [1]). However, there were no data on mutations underlying genetic disorders in Ku-

KARGER

Fax +41 61 306 12 34
E-Mail karger@karger.ch
www.karger.com

1422–2795/05/0081–0040$22.00/0

Accessible online at:
www.karger.com/cmg

Elena Samilchuk, MD, PhD, DMSc
E-Mail samilchuk@hotmail.com

Table 1. Mutations found in patients from Kuwait

Disease	Direct tests for common mutations or those previously reported in Arabian patients	Mutation analysis of coding exons and splice junctions
CF and CBAVD	delF508, S549N, W1282X, 1548delG, I1234V, 3120+1G→A, IVS8-polyT	I506T, C128R
PKU	IVS10-11G→A (Mediterranean)	1089delG, IVS4+5G→T, IVS2+5G→C, 1055delG, I224T, E280K, P281L, DelE2/IVS2nt1
SMA	Homozygous deletion of SMN gene (exons 7 and 8)	
G6PD deficiency	563C→T (Mediterranean) 202G→A (African)	1003G→A (Chatham) 143T→C (Aures)
CAII deficiency	IVS2+1G→A (Arabic)	
Tay-Sachs disease		R504H
CPEO	4.9del	
ASL deficiency	E241K	
LHON	11778G→A	
FMF	M694V	

ASL = Arginine succinate lyase; CAII = carbonic anhydrase II; CPEO = chronic progressive external ophthalmoplegia; FMF = familial Mediterranean fever; LHON = Leber hereditary optic neuropathy.

wait, and a molecular genetic diagnostic service did not exist. In the early 1990s, the Kuwaiti Ministry of Health took the decision to establish a DNA-diagnostic laboratory in the Kuwait Medical Genetics Center, and the author pioneered the introduction of genetic molecular diagnostic services in Kuwait. This paper summarizes the author's major findings in the field of single gene disorders, as well as problems hampering this type of genetic service in non-European populations.

Subjects and Methods

DNA samples from probands and other members of the families with single gene disorders were collected between 1993 and 2003 by extraction from peripheral blood cells using the 'salting out' technique [2]. All tests were based on polymerase chain reaction (PCR) amplification using different types of primers (flanking mutation sites, allele-specific sites or with mismatches creating a new restriction site). Most of the direct tests for mutations were based on the detection of restriction fragment length polymorphisms within PCR-amplified fragments, which discriminated between normal and mutant sequences. The sizing of intact or digested PCR fragments was carried out by agarose or polyacrylamide gel electrophoresis. Direct testing for mutations was also carried out by sequencing of PCR-amplified fragments using an ABI dye terminator kit and 310 ABI Genetic Analyzer. In the case of cystic fibrosis (CF) these 'in-house' tests have been recently supplemented by the ABI CF Mutation Kit (PCR, OLA, and fragment sizing with the ABI Genetic Analyzer). Some samples negative for common mutations were further studied by heteroduplex analysis in MDE gels, and sequencing of coding exons and associated splice junctions. All tests were provided to the families free of charge.

Results and Discussion

The direct testing for mutations reported to be common elsewhere was the major approach. Further study of negative samples (heteroduplex analysis and sequencing of coding exons with splice junctions) was conducted only for certain disorders since there was no funding for this type of research. Direct testing for common mutations had variable success for different genetic disorders (table 1). In the case of CF, only three mutations (delF508, G542N and W1282X) were identified by using the commercial kit. These mutations accounted for 34% of all CF alleles in Kuwait and 55% of Arabic alleles. As several other mutations (1548delG, I1234V, 3120+1G→A and H139L) were previously reported in Saudi Arabia [3, 4], corresponding tests were designed. Three of these mutations (1548delG, I1234V and 3120+1G→A) were present in Kuwaiti patients, and together they accounted for 24% of the Arab CF alleles from Kuwait. Thus, testing for these mutations in Arab CF patients is strongly indicated. Further mutation analysis of selected CFTR exons revealed one novel (C128R) and one rare (I560T) mutation. However, 38% of CF alleles in Kuwait have yet to be identified. In patients with congenital bilateral absence of the vas deferens (CBAVD), delF508 was found in 17% and IVS8-5T in 33% of Arabic CBAVD alleles. This unusually high frequency of IVS8-5T seems to be characteristic of Arab CBAVD patients [5].

Hemochromatosis (HH) is one of the most common genetic disorders in Europeans and seems to be rare in Kuwait. The major HH mutation (C282Y) was absent in 5 patients with a provisional diagnosis of HH as well as in 162 healthy Kuwaitis, suggesting that the population frequency of C282Y is less than 0.0031 (much lower than the population frequencies of 0.02–0.1 reported for Europeans). Thus, testing for C282Y (one of the most frequently used tests in European DNA-diagnostic laboratories) seems to be of little importance in Kuwait.

In contrast, spinal muscular atrophy (SMA) is common in Kuwait. All SMA patients had a homozygous deletion of the SMN1 gene (both exons 7 and 8 were missing in all patients). Many SMA patients had also a homozygous deletion of the NAIP gene. Such dual deletion of NAIP and SMN1 genes seems to be more frequent in Arab patients with SMA than in Europeans: 85 versus 45% in SMA type I and 40 versus 17% in types II/III [6].

Phenylketonuria (PKU) is the most common metabolic single gene disorder in Kuwait. The direct testing for four common European mutations (R408W, IVS12+1G→A, R261Q, and IVS10-11G→A) was successful only for the latter mutation (34.6% of PKU alleles), and further analysis of coding exons and splice junctions revealed eight other mutations. Two of these mutations (DelE2-IVS2nt1 and I224T) seem to be of Arab origin. Indeed, in addition to the Kuwaiti Arabs in our study, these two mutations were earlier reported only in Palestinians and Egyptians [7].

A rare condition, rhabdomyolysis, is caused by mutations in the carnitine palmitoyltransferase II gene, and a common mutation (S113L) has been reported [8]. However, this mutation was absent in a Kuwaiti family with several affected sibs and because it was just a single family, further mutation analysis unfortunately was not performed.

Osteopetrosis, another rare disease with renal tubular acidosis and cerebral calcification, is mostly confined to Arabs. The disease is caused by a deficiency in the carbonic anhydrase II gene, and a unique Arabic mutation (IVS2+1G→A) has been reported [9]. The direct test for IVS2+1G→A revealed this mutation in Kuwaiti patients [10].

The E241K mutation was identified in a Kuwaiti proband with arginine succinate lyase deficiency in the laboratory of Prof. Koch in Germany [11]. Since the mutation was known, it was possible to test more than 20 family members who asked for carrier testing as part of premarital counseling.

Three Kuwaiti families with Tay-Sachs disease have been studied. The lack of common mutations among non-Jewish patients with the disease made the sequencing of coding exons of the HEXA gene the only option for the analysis. The R504H mutation was identified in one family (affected sibs were homozygous). This mutation has only been reported in 3 patients (a Lebanese, an Assyrian and an American, all from consanguineous families) to date, thus suggesting a middle-eastern origin of R504H. No mutations, including pseudodeficient variants, were found in two other Kuwaiti families although all coding exons and splice junctions have been sequenced. In these cases, the mutation is possibly located outside the coding region, however incorrect enzymatic diagnosis of Tay-Sachs disease should also be considered. Indeed, heat-labile mutations in the related HEXB gene (such as 11672 G→A reported in Palestinian Arabs and Jews) may lead to erroneous enzymatic genotyping regarding Tay-Sachs disease [12].

Familial Mediterranean fever is very rare among Kuwaiti nationals, but it had been reported to be common among Jordanian and Palestinians residing in Kuwait in the 1980s [1]. Although direct tests for four common mutations in exon 10 as well as sequencing of this fragment have been offered since 1998, only one family (Jordanians) has been referred for testing. Both affected sibs were homozygous for M694V.

A number of samples were referred for mitochondrial DNA (mtDNA) mutations (MELAS 3243A→G, MERRF 8344A→G and NARP/Leigh disease 8993T→C/G), but on testing all were negative for these mutations. Among patients with Leber hereditary optic neuropathy only one of four mutations tested for the disease (3460G→A, 11778 G→A, 14484T→C and 14459G→A) was identified. In a single patient with chronic progressive external ophthalmoplegia, the common 4.9-kb mtDNA deletion was detected using a PCR test with flanking primers.

Glucose-6-phosphate dehydrogenase (G6PD) deficiency is common in Kuwait (6.5% of Kuwaiti males). Direct testing for common Mediterranean and African mutations as well as sequence analysis of several negative cases showed that G6PD deficiency in Kuwait was caused by four mutations (563C→T, 202G→A, 1003G→A and 143T→C) [13].

The majority of the families studied were consanguineous; usually the parents were first cousins. As a result, homozygosity by descent was observed in most cases of autosomal recessive disorders. For example, 88% of CF and 77% of PKU probands had identical mutations on maternal and paternal chromosomes inherited from one of the grandparents.

Among mutations identified in Kuwait, there were Mediterranean (IVS10-11G→A in PKU, 563C→T in G6PD and M694V in familial Mediterranean fever), African (3120+1G→A in CF and 202G→A in G6PD) and Arabic (1548delG and I1234V in CF, IVS2+1G→A in carbonic anhydrase II and DelE2-IVS2nt1 and I224T in PKU) variants.

In conclusion, more efforts aimed at the identification of mutations underlying genetic disorders in Kuwait as well as in other Gulf countries are warranted. This can be achieved by focusing genetic research within the academic institutions of Gulf countries on this goal.

References

1 Teebi AS: Autosomal recessive disorders among Arabs: An overview from Kuwait. J Med Genet 1994;31:224–233.
2 Miller S, Dykes D, Polesky H: A simple salting out procedure for extraction DNA from human nucleated cells. Nucleic Acids Res 1988; 16:1215.
3 el-Harith EA, Dork T, Stuhrmann M, Abu-Srair H, al-Shahri A, Keller KM, Lentze MJ, Schmidtke J: Novel and characteristic CFTR mutations in Saudi Arab children with severe cystic fibrosis. J Med Genet 1997;34:996–999.
4 Banjar H, Kambouris M, Meyer BF, al-Mehaidib A, Mogarri I: Geographic distribution of cystic fibrosis transmembrane regulator gene mutations in Saudi Arabia. Ann Trop Paediatr 1999;19:69–73.
5 Lissens W, Mahmoud KZ, El-Gindi E, Abdel-Sattar A, Seneca S, Van Steirteghem A, Liebaers I: Molecular analysis of the cystic fibrosis gene reveals a high frequency of the intron 8 splice variant 5T in Egyptian males with congenital bilateral absence of the vas deferens. Mol Hum Reprod 1999;5:10–13.
6 Samilchuk E, D'Souza B, Bastaki L, al-Awadi S: Deletion analysis of the SMN and NAIP genes in Kuwaiti patients with spinal muscular atrophy. Hum Genet 1996;98:524–527.
7 Scriver CR, Hurtubise M, Konecki D, Phommarinh M, Prevost L, Erlandsen H, Stevens R, Waters PJ, Ryan S, McDonald D, Sarkissian C: PAHdb 2003: What a locus-specific knowledge base can do. Hum Mutat 2003;21:333–344.
8 Taroni F, Verderio E, Dworzak F, Willems PJ, Cavadini P, DiDonato S: Identification of a common mutation in the carnitine palmitoyltransferase II gene in familial recurrent myoglobinuria patients. Nat Genet 1993;4:314–320.
9 Fathallah DM, Bejaoui M, Sly WS, Lakhoua R, Dellagi K: A unique mutation underlying carbonic anhydrase II deficiency syndrome in patients of Arab descent. Hum Genet 1994;94: 581–582.
10 Samilchuk E, D'Souza B, Al-Awadi S: A Kuwaiti patient with osteopetrosis, renal tubular acidosis and cerebral calcification is homozygous for a splice junction mutation in the carbonic anhydrase gene. Med Principles Pract 1996;5:234–237.
11 Linnebank M, Tschiedel E, Haberle J, Linnebank A, Willenbring H, Kleijer WJ, Koch HG: Argininosuccinate lyase (ASL) deficiency: Mutation analysis in 27 patients and a completed structure of the human ASL gene. Hum Genet 2002;111:350–359.
12 Narkis G, Adam A, Jaber L, Pennybacker M, Proia RL, Navon R: Molecular basis of heat labile hexosaminidase B among Jews and Arabs. Hum Mutat 1997;10:424–429.
13 Samilchuk E, Al-Suliman I, Usanga E, Al Awadi S: 1: Glucose-6-phosphate dehydrogenase (G6PD) mutations and UDP-glucuronosyltransferase promoter polymorphism among G6PD deficient Kuwaitis. Blood Cells Mol Dis 2003;31:201–205.

Community Genetics

Community Genet 2005;8:44–47
DOI: 10.1159/000083338

Recently Available Techniques Applicable to Genetic Problems in the Middle East

Pinar T. Ozand[a] Ali Al Odaib[a] Nadia Sakati[a] Ali M. Al-Hellani[b]

[a]Department of Genetics and [b]Section of Preimplantation Genetic Diagnosis, In vitro Fertilization Unit, King Faisal Specialist Hospital and Research Center, Riyadh, Saudi Arabia

Key Words
Genetic diseases · Inherited metabolic diseases · Middle East

Abstract
In this paper, we address the preventive health aspects of genetic problems in the Middle East and provide guidelines to prioritize preventive strategies. Applications of various novel genetic techniques such as comprehensive neonatal screening, high throughput heterozygote detection, preimplantation genetic diagnosis, Affymetrix systems, the NanoChip system and a new way of sensitive karyotyping for single-cell chromosome abnormalities are discussed. In conclusion, from the various genetic techniques available, each country should adopt strategies most suitable to its genetic needs and should prioritize the programs to be used in prevention.

Introduction

The Arabian Peninsula, particularly Saudi Arabia, holds a special place in the world as far as inherited diseases are concerned. Inherited neurometabolic diseases prevalent in Saudi Arabia are particularly important since they usually lead to morbidity requiring chronic care and to mortality. The high rate of consanguinity and tribal marriages is responsible for an inordinately large number of such diseases. In Saudi Arabia, cousin marriages account for about 60%, with half involving first cousins and one fourth involving double first cousin marriages [1].

Data derived from the author's experience at the Metabolic Clinics of the King Faisal Specialist Hospital and Research Center (KFSH & RC) are discussed for the period from 1995 until 2000. The novel technologies reviewed in this paper are those currently available in the Department of Genetics at KFSH & RC. Some of the major neurometabolic diseases encountered at one of the author's clinic at KFSH & RC within a 5-year period are shown in table 1.

In the face of such a public health problem that afflicts a significant segment of the population requiring chronic care, the most important policy decision is to design a preventive health system to combat the problem.

The most important chore of a geneticist involved in preventive medicine is to prioritize the genetic problems of the country and design programs to address them. The geneticist must also select the technique that would offer a solution to the problem.

KARGER

Fax +41 61 306 12 34
E-Mail karger@karger.ch
www.karger.com

1422–2795/05/0081–0044$22.00/0

Accessible online at:
www.karger.com/cmg

Pinar T. Ozand, MD, PhD
Department of Genetics, King Faisal Specialist Hospital and Research Center
PO Box 3354
Riyadh 11211 (Saudi Arabia)
Tel. +966 1 4424263, Fax +966 1 4424585, E-Mail Ozand@kfshrc.edu.sa

Table 1. Some of the common inherited disorders encountered during a 5-year period (1995–2000) in one metabolic clinic at the King Faisal Specialist Hospital and Research Center

Diseases	Number seen
Primary lactic acidosis and mitochondriopathies	147
Methylmalonic acidemia	75
Maple syrup urine disease	61
Propionic acidemia	55
Neuronal ceroid lipofuscinosis	45
Classic PKU	42
Patients within autism spectrum	41
Niemann-Pick disease type B, aggressive type	39
Classic homocystinuria	39
6-Pyruvoyltetrahydropterin-synthase-deficient PKU	30
Glutaric aciduria type 1	29
Biotinidase deficiency	26
3-Hydroxy-3-methylglutaryl CoA lyase deficiency	23
Morquio's disease	22
Argininosuccinic aciduria	21
PGAL transferase-deficient galactosemia	15
Gaucher's disease	15
Citrullinemia	11
Very-long-chain acyl CoA dehydrogenase deficiency	10
Medium-chain acyl CoA dehydrogenase deficiency	10

PKU = Phenylketonuria; PGAL = galactose-1-phosphate uridyl transferase.

Genetic Priorities

What and how to prioritize? The following list may help to prioritize genetic diseases: (1) diseases that are encountered frequently; (2) genetic diseases that have known diagnostic procedures and intervention strategies should always be considered prior to others; (3) genetic diseases that have no known management procedures but that could be prevented; (4) chronic diseases that are not necessarily untreatable but disrupt family structure and function; (5) treatable diseases where early identification is important, and (6) diseases that may be rare but incur huge expenditure in terms of health costs.

Novel Genetic Diagnostic Techniques for Disease Prevention

The diagnosis of a genetic disease depends upon (1) a biochemical marker, (2) a molecular genetic marker or (3) chromosomal abnormality. Some examples for the employment of such markers are as follows:

Neonatal Screening

Powerful biochemical techniques are available for the identification of a metabolic disease. Some examples are: gas chromatography/mass spectrometry and tandem mass spectrometry. The latter is a high throughput method. From the perspective of preventive genetics, tandem mass spectrometry can be used to identify organic and amino acidemias soon after birth thereby permitting management of a large number of patients [2]. Other biochemical methodologies for screening biotinidase deficiency, congenital adrenal hyperplasia, galactosemia, and hypothyroidism are also available. These diseases, once identified, can successfully be managed. However, none of these techniques are suitable for carrier detection.

Molecular Genetic Markers

Genetic disorders can be divided into single-gene disorders or polygenic diseases. In the case of single-gene disorders, there are high throughput reliable techniques for molecular genetic markers. From a preventive point of view, their value is more for population screening and family planning.

DNA Papers

In order to use molecular genetic techniques for screening and intervention, the mutation in the index patient must be established. With the recent techniques available, this can be achieved rapidly by the use of a multiple column sequencer, such as a Megabace. Once the mutation is identified, there are high throughput techniques that can screen the entire extended family to identify the carriers. It is also possible to collect a specimen of blood for screening on filter paper [3], which can then be mailed to the testing center. The DNA on this paper is stable up to 7 years at room temperature.

Heterozygote Detection

This is important for preventive genetics in that the family members can be counseled for marriages. Without disturbing the tradition of consanguineous marriages, in cousins who want to marry but are worried about the consequences, genetic diseases can be identified. The carrier detection can be done quickly by a light-cycler or by a NanoChip system [4] and need not always include gene sequencing. This information can also be used later on for preimplantation genetic diagnosis (PGD) assuring a normal baby in the next pregnancy.

DNA Microarrays for Karyotyping

A recent technology for karyotyping permits the identification of even a very small abnormality in a chromosome [5]. It is based on BAC-clone-embedded DNA microarrays, and it is much more sensitive compared to conventional techniques such as microscopic karyotyping, fluorescence in situ hybridization or chromosome painting. It can rapidly identify chromosome abnormalities missed by microscope-based techniques. It is based on complete human genome hybridization [6]. Together with the use of the newly available enzymes that augment the amount of DNA by using degenerate primers, this technique can now be applied on single cells [7]. We have recently been able to use this technique with success to identify the chromosomal abnormality in a single cell [8]. Previous complete human genome hybridization procedures were time consuming as they required lengthy procedures and were therefore not suitable for single-cell application [9]. The immediate preventive genetic application of this method is PGD [10] in families who are at high risk for various chromosomal disorders that are encountered in habitual miscarriages. It is also an important tool for the diagnosis of dysmorphic syndromes.

Detection of PGD Mutations in Preventive Genetics

In this technique, a single cell is biopsied from an eight-cell-stage embryo [10]. The cell removed is used to identify a mutation. The normal embryo is then placed into the uterus. At present, the technique is perfected so that there are practically no false-negative or -positive results. The PGD is most suitable for countries where therapeutic abortions based on prenatal diagnosis are not permitted for religious and social reasons.

Some examples for the use of these prevention techniques are as follows. For example in Saudi Arabia, propionic acidemia is a very common disorder. One particular tribe has a very severe variant of propionic acidurias and it causes severe morbidity despite all available appropriate management procedures. Despite intensive and chronic care in these patients, the results are not always rewarding. This variant of propionic aciduria was found to be caused by one mutation. We have recently used PGD successfully in 2 families.

Several genetic diseases could successfully be managed by conventional methods. Examples are homocystinuria, 6-pyruvoyltetrahydropterin-synthase-deficient phenylketonuria, Niemann-Pick's disease and Gaucher's disease [11–13]. The 6-pyruvoyltetrahydropterin-synthase-deficient phenylketonuria and homocystinuria cause serious morbidity when left untreated. On the other hand, affected children on medications grow to become healthy adults. However, the medications used in their treatment are prohibitively expensive, amounting to $ 30–40,000 per year. The enzyme treatment of Niemann-Pick and Gaucher's diseases are also in the order of $ 30–40,000 per year for a child but becomes even more expensive when they become adults. These four diseases are lifelong conditions and at any point in their lives when the medications are not available, the patient suffers serious neurological debilitation. Therefore, PGD and premarital and extended family screening for these disorders are excellent preventive genetic goals for Saudi Arabia.

Similarly, these preventive methods can be used for diseases that have absolutely no treatment procedures available at present. Examples are the Zellweger syndrome and Canavan's disease.

Affymetrix Systems to Study Genetic Diseases

The genetic techniques to study polygenic disease are at their infancy. Initially homozygosity mapping was the method of choice. This technique tries to establish co-inheritance of a genetic marker with a disease [14]. If two characters are always inherited together it means the location of that marker should be in close proximity of the disease gene. Then candidate genes in the region are sequenced for polymorphisms and mutations. More recently, an mRNA-oligonucleotide-based chip system, the Affymetrix system [15], has superseded this technique. Using this method, the 'genetic signature' of the disease is secured. The altered gene functions of many genes are established simultaneously in order to obtain a more comprehensive biochemical picture.

Pharmacogenomics

The more recent NanoChip system has been an important contribution to pharmacogenomics. In this field, single-nucleotide polymorphic markers are studied in relation to the patient's tolerance of conventional drugs [4]. It is an area that will receive significant attention in the near future. The polymorphic markers are also useful in defining populations and ethnicity [16]. This will have implications for preventive genetics enabling to identify the drug of choice for a disease or to ascertain the prevalence of a particular disease in an ethnic group.

Conclusion

Some of the novel genetic techniques described in this paper are most valuable for preventive medicine. With the availability of such high-throughput and mass-applicable techniques, each country should establish a genetic facility to suit the genetic needs of its own community and population. Despite a crossover of population among countries of the Arabian Peninsula, each country has its own genetic problem that needs different preventive genetic approaches. These problems should be prioritized for each country, and preventive genetic programs should be initiated with one or several of the aforementioned techniques.

References

1 al-Abdulkareem AA, Ballal SG: Consanguineous marriage in an urban area of Saudi Arabia: Rates and adverse health effects on the offspring. J Community Health 1998;23:75–83.

2 Rashed MS, Rahbeeni Z, Ozand PT: Application of electrospray tandem mass spectrometry to neonatal screening. Semin Perinatol 1999; 23:183–193.

3 McCabe ER: Utility of PCR for DNA analysis from dried blood spots on filter paper blotters. PCR Methods Appl 1991;1:99–106.

4 Evans JG, Lee-Tataseo C: Determination of the factor V Leiden single-nucleotide polymorphism in a commercial laboratory by use of NanoChip microelectronic array technology. Clin Chem 2002;48:1406–1411.

5 Gunn S, Mohammed MS, Reveles X, Viskochil D, Palumbos J, Johnson-Pais T, Hale D, Lancaster J, Hardies J, Bioesplug-Tanguy O, Leach R: Characterization of a patient with central system dysmyelination and cryptic unbalanced translocation between chromosome 4q and 18q. Am J Med Genet 2003;120A:127–135.

6 Forozan F, Karhu R, Kononen J, Kallioniemi A, Kallioniemi OP: Genome screening by comparative genomic hybridization. Trends Genet 1997;13:405–409.

7 Preston GM: Cloning gene family members using PCR with degenerate oligonucleotide primers. Methods Mol Biol 2003; 226:485–498.

8 Hellani A, Coskun S, Benkhalifa M, Tbakhi A, Sakati N, Al-Odaib A, Ozand P: Multiple displacement amplification on single cell and possible PGD applications. Mol Hum Reprod 2004;10:847–852.

9 Wells D, Delhanty JD: Comprehensive chromosomal analysis of human preimplantation embryos using whole genome amplification and single cell comparative genomic hybridization. Mol Hum Reprod 2000;6:1055–1062.

10 Verlinsky Y, Kuliev A: An Atlas of Preimplantation Diagnosis. Pearl River, Parthenon, 2000.

11 Weinreb NJ, Charrow J, Andersson HC, Kaplan P, Kolodny EH, Mistry P, Pastores G, Rosenbloom BE, Scott CR, Wappner RS, Zimran A: Effectiveness of enzyme replacement therapy in 1028 patients with type 1 Gaucher disease after 2 to 5 years of treatment: A report from the Gaucher Registry. Am J Med 2002; 113:112–119.

12 Al Aqeel A, Ozand PT, Gascon GG, Hughes H, Reynolds CT, Subramanyam SB: Response of 6-pyruvoyl-tetrahydropterin synthase deficiency to tetrahydrobiopterin. J Child Neurol 1992;7(suppl):S26–S30.

13 Wilcken DE, Wilcken B, Dudman NP, Tyrrell PA: Homocystinuria – The effects of betaine in the treatment of patients not responsive to pyridoxine. N Engl J Med 1983;309:448–453.

14 Turner M, Barnby G, Bailey A: Genetic clues to the biological basis of autism. Mol Med Today 2000;6:238–244.

15 Liu G, Loraine AE, Shigeta R, Cline M, Cheng J, Valmeekam V, Sun S, Kulp D, Siani-Rose MA: NetAffx: Affymetrix probesets and annotations. Nucleic Acids Res 2003;31:82–86.

16 McCarthy JJ, Hilfiker R: The use of single-nucleotide polymorphism maps in pharmacogenomics. Nat Biotechnol 2000;18:505–508.

Community Genetics

Community Genet 2005;8:48–51
DOI: 10.1159/000083339

Attitudes toward Genetic Counseling in the United Arab Emirates

L.I. Al-Gazali

Department of Pediatrics, Faculty of Medicine and Health Sciences, United Arab Emirates University, Al Ain, United Arab Emirates

Key Words
Consanguinity · Genetic counseling · Prenatal diagnosis · United Arab Emirates

Abstract
Aims: To assess the level of understanding of genetic advice given in the Genetic Clinic and attitudes toward consanguineous marriages, and prenatal, abortion and preconception diagnoses. ***Methods:*** One hundred couples underwent structured interviews, and various social and educational data, reason for referral and diagnosis and advice given were recorded. Three months later, the couples were asked open-ended questions about the perceived causation of the disease, recurrence risk, plans for births, and prenatal, abortion and preconception diagnoses. ***Results:*** Half of the couples acknowledged a genetic basis for their child's condition but only 10 remembered the risk given to them. There was a high correlation between educational level and remembering the risk, and the number of healthy children and future plans for further children. Almost half preferred consanguineous marriages and only 10% agreed with prenatal diagnosis and abortion, while 75% agreed with carrier screening and preconception diagnosis in affected families. ***Conclusion:*** Effective genetic counseling in this community requires an informed educated population and introduction of carrier screening and preconception diagnosis in affected families.

Introduction

The population of the United Arab Emirates (UAE) is a mixture of different ethnic groups, and only 20% are UAE nationals. The family size is large, and women continue to reproduce well after the age of 40 years. Currently, there are no preventive programs and abortion is not legal. Consanguineous marriages are very common in this population (50.4%), particularly first cousin marriages (26.0%) [1]. Several studies from the UAE have shown a high frequency of genetic disorders, particularly autosomal recessive disorders [2–4]. These disorders are mostly of rare and complex types, with a high frequency of new syndromes. There is also clustering of certain disorders or syndromes in specific tribes or families. Clinical experience indicates that genetic counseling has not been very effective for this population.

The aim of this study was therefore to assess the level of understanding of the genetic advice given to the couple in the Genetic Clinic, and attitudes toward consanguineous marriages, and prenatal, abortion and preconception diagnoses.

Subjects and Methods

The study population comprised couples seen in the Genetic Clinics in Al Ain and Tawam hospitals, Al Ain, between September 2000 and September 2002. In 6 cases, only 1 spouse was available. All spouses underwent a semi-structured interview by the author, and various personal, social and educational data were obtained.

KARGER

Fax +41 61 306 12 34
E-Mail karger@karger.ch
www.karger.com

1422–2795/05/0081–0048$22.00/0

Accessible online at:
www.karger.com/cmg

Prof. L.I. Al-Gazali, MB ChB, MSc, DCH, FRCP, FRCPCH
Department of Pediatrics, Faculty of Medicine and Health Sciences
United Arab Emirates University, PO Box 17666
Al Ain (United Arab Emirates), Tel. +971 3 7039415, Fax +971 3 7039668
E-Mail l.algazali@uaeu.ac.al or algazali@hotmail.com

Table 1. Types of disorders in the study population

Inheritance	n
Autosomal recessive	78
Autosomal dominant	7
X-linked recessive	5
Chromosomal	10
Total	100

Table 2. Reproductive plans, current family size and affected births

Healthy births (p = 0.023)			Affected births (p = 0.350)		Total
wish more children	≤5	>5	no death	1 or more death	
Yes	26	10	21	15	36
No	15	20	25	10	35
Up to God	20	9	16	13	29
Total	61	39	62	38	100

The reason for referral, diagnosis and advice given was then recorded. No explanatory letter was given after the counseling session and no follow-up visit by a genetic counselor was provided (it was not available). With one exception the diagnosis in the children was not established at the time of referral and therefore the reason for referral was primarily to establish the diagnosis. In one family the reason for referral was premarital genetic counseling, due to the family history of an autosomal recessive disorder.

A total of 100 families were recruited for the study. Of these, 90 were consanguineous, 7 had complex consanguinity, 2 were double cousins, 67 were first cousins, 9 first cousins once removed, and 5 were second cousins. The coefficient of inbreeding in the study group was 0.0526 compared to 0.0222 in the general population. Three quarters (75) of the families had 1–2 affected children, and 25 had 2–7 affected children. One or more death and up to 5 had been experienced by 38 families. In 78 families the disorder was autosomal recessive, in 7 autosomal dominant, in 5 X-linked recessive and in 10 it was chromosomal (table 1).

The couples were again interviewed 3 months later and were asked open-ended questions regarding the perceived causation of the disease, the recurrence risk given to them, their plans for births, marriages, and prenatal diagnosis, abortion and preconception diagnosis. Parental responses in acknowledging heredity and reproduction plans were subsequently evaluated against educational level, family size and the number of affected children using χ^2 tests and logistic regression analysis.

A sub-sample of 20 families were further reassessed 2 years after the initial counseling.

Results

Almost all parents attributed their child/children's illness to God's will. However, when pressurized to give other possible explanations for the disorder, 52 acknowledged a genetic basis for their child's condition without truly accepting or understanding it. There was a significant correlation between the education of the couple and accepting genetic counseling. (χ^2 test: p = 0.006 and p = 0.000 for wives and husbands, and logistic regression analysis p = 0.004 and p = 0.000 for wives and husbands, respectively). Most parents did not accept the inheritance of disease unless the family history made this obvious. Only 10 couples remembered the risk given to them. In each of these cases, the husband had a university degree and in 5 both the husband and wife had a university degree. However, those who had more than 1 affected child repeatedly questioned the risk given to them. This indicates that the parents did not actually understand the genetic mechanism and the mathematical probabilities explained to them. Several stated that 'they did not know' the cause of the disorder. A total of 40 couples interviewed confessed to resorting to traditional methods of healing by taking their children to local healers. Several children examined subsequently had cautery marks on their body. This happened even when the parents were educated at university level, due to grandparents' insistence.

When asked if they would prefer for themselves and for their children to marry within the family, 45 responded that cousin marriages were preferred, 33 rejected such marriages and 22 were not sure. Most parents who preferred cousin marriages attributed their preference to the traditional way of thinking that 'a cousin takes better care of you'.

Attitudes towards future births were also assessed. A third of the couples (n = 36) said they would like to have more children but that depended on God's will, 35 did not want to have more children, and 29 said that it was not in their hands but in 'God's hand'. There was a significant correlation between the number of healthy children in the family and plans to have more children (table 2). However, there was no correlation between the number of affected children and future plans.

When asked about prenatal diagnosis, 40 couples said they would have prenatal diagnosis if it were available, 45 said they would not, and 15 were not sure. However, of the 40 couples who agreed with prenatal diagnosis only

10 said they would have an abortion if it were available at 10 weeks. The remaining 30 couples were either against abortion (n = 26) or not sure (n = 4).

Twenty families were seen 2 years after the initial counseling. Eighteen had gone ahead with a pregnancy. Eight families had another affected child, 3 families had more than 1 affected birth, 5 had normal births and 1 was still pregnant. One family who came initially for premarital counseling due to the family history of an autosomal recessive condition went ahead with marriage to a first cousin in spite of the risk presented to them (2 brothers married 2 sisters) and had affected twin girls. Two families refrained from having more children. Both families had 3 affected children, with 1 normal in 1 family and no normal children in the other family. In the former family, the normal child got married to her first cousin, even though 2 years earlier the parents were against cousin marriage. Three families asked for prenatal diagnosis, in 1 it was not available for the condition. The second family's prenatal diagnosis by ultrasound revealed an affected pregnancy at 18 weeks of gestation, but the family was unable to obtain an abortion for legal reasons. The third family went abroad for prenatal diagnosis and abortion of the fetus if affected.

Discussion

Several studies have shown that genetic information, particularly the risk of recurrence of the disorder in future children, is often misunderstood or incompletely understood by families affected with a genetic disorder [5–11]. The difficulty is usually in grasping in statistical terms the risk of having normal or affected children. It has also been shown that education and familiarity with the disorder are important factors in understanding the genetic advice given.

Almost all parents in this study attributed their child/children's illness to God's will. Those who were highly educated or those whose family history made inheritance obvious acknowledged the genetic basis of their child's condition. However, only a few remembered the risk given to them (n = 10), and within this group those who had affected children repeatedly questioned that risk. This indicates that even those who acknowledged the genetic basis and remembered the risk did not actually understand the genetic mechanisms and the mathematical probabilities explained to them. In particular, parents found it difficult to accept the inability to specify why a particular child was spared while another was affected. Therefore, parents preferred to attribute the reason for individual suffering to God's will rather than to chance. Such belief helps to alleviate feelings of guilt by relating a child's problems to God's purposeful action rather than a 'blind' probabilistic event [12]. In addition, several local beliefs, which are transmitted over generations, also affect people's response to genetic disorders. These are very common in the UAE. A large number of couples (n = 40) interviewed in this study admitted resorting to non-medical native healers for explanation and relief. These healers use rituals containing a lot of religion-based supportive psychotherapy. Some however, resort to cauterization as a form of therapy for genetic disorders. Many children seen in this study had cautery marks 3 months after the initial counseling session.

Attitudes toward consanguinity were also assessed. Nearly half of the couples interviewed (45) favored these marriages for themselves and their children. Only 33 were against such marriages and 22 were not sure. Even among the families who rejected consanguineous marriages, 2 went ahead with such a marriage 2 years after the initial counseling. This is not surprising as consanguinity is deeply rooted in the Arab culture, predating Islam, and thus has been practiced over many generations. In fact, a recent study from the UAE showed that consanguineous marriages are increasing rather than decreasing [1]. Therefore, counseling may have very little impact, as marriage patterns can be very resistant to change [13]. In addition, several studies showed that genetic counseling does not influence planning the birth of another child in affected families [5, 6, 12]. The perceived burden of having another affected birth, family size and prior reproduction plans are far more likely to influence reproductive behavior [5, 6, 12]. In this study a strong correlation between the number of healthy children, rather than the number of affected children, and plans to have more children was found. This is particularly true for the Arab Bedouin community where married women are expected to produce a number of healthy children, in particular sons. Therefore, the couple usually continues reproducing until an accepted number of healthy children, especially sons, is achieved. In such a setting, and particularly when there are no options available to couples, genetic counseling is not expected to have a major impact on the frequency of genetic disorders. For counseling to be effective in the UAE, options which are acceptable to the community should be available for the couples. Most couples interviewed in this study did not favor prenatal diagnosis and the abortion of affected fetuses even undertaken at 10 weeks of gestation. However, the majority agreed with preconception diagnosis. Therefore, genetic counseling offering premarital screening of carriers in specific families and preconception diagnosis is the right approach for the prevention of genetic diseases in this community. In addition, there should be public educational programs in genetics, trained genetic counselors from the community and equipped laboratories with trained personnel. Explanatory letters after counseling should also be introduced as a mean to improve the understanding of genetic counseling in this community.

References

1 Al-Gazali LI, Bener A, Abdulrazzaq Y, Micallef R, Al Kayat AI, Gaber T: Consanguineous marriages in the UAE. J Biosoc Sci 1997;29:491–497.

2 Al-Gazali LI, Dawodu A, Sabarinath K, Varghese M. The profile of congenital abnormalities in the UAE population. J Med Genet 1995;32:7–13.

3 Al-Gazali Li, Sztriha L, Dawodu A, Bakir M, Varghese M, Varady E, Scorer J, Abdulrazzaq YM, Bener A, Padmanabhan R: Pattern of central nervous system anomalies in a population with a high rate of consanguineous marriages. Clin Genet 1999;55:95–102.

4 Al-Gazali Li, Bakir M, Hamid Z, Varady E, Varghese M, Haas D, Bener A, Padmanahban R, Abdulrazzaq YM, Dawodu A: Birth prevalence and pattern of osteochondrodysplasias in an inbred high risk population. Birth Defects Res Part A Clin Mol Teratol 2003;67:125–132.

5 Panter-Brick C: Parental responses to consanguinity and genetic disease in Saudi Arabia. Soc Sci Med 1991;33:1295–1302.

6 Emery AE, Raeburn JA, Skinner R, Holloway S, Lewis P: Prospective study of genetic counseling. Br Med J 1979;i:1253–1256.

7 Ever-Keibooms G, Van Den Berghe H: Impact of genetic counseling: A review of published follow-up studies. Clin Genet 1979;15:465–474.

8 Kessler S, Levine E: Psychological aspects of genetic counseling. IV. The subjective assessment of probability. Am J Med Genet 1987;28:361–370.

9 Sorenson JR, Scotch NA, Swazey JP: Reproductive plans for genetic counseling clients not eligible for prenatal diagnosis. Am J Med Genet 1987;28:345–352.

10 Chase GA, Faden RR, Holtzman NA, Chwalow J, Leonard CO, Lopes C, Quaid K: Assessment of risk by pregnant women: Implication for genetic counseling and education. Soc Biol 1986;33:57–64.

11 Wertz DC, Sorenson JR, Heeren TC: Client's interpretation of risks provided in genetic counseling. Am J Hum Genet 1986;39:253–264.

12 Raz AE, Atar M, Rodnay M, Shoham-Vardi I: Between acculturation and ambivalence: Knowledge of genetics and attitudes towards genetic testing in a consanguineous Bedouin community. Community Genet 2003;6:88–95.

13 Shiloh S, Reznik H, Bat-Miriam-Ktznelson M, Goldman B: Pre-marital genetic counselling to consanguineous couples: Attitudes, beliefs and decisions among counselled, noncounselled and unrelated couples in Israel. Soc Sci Med 1995;41:1301–1310.

Community Genetics

Community Genet 2005;8:52–55
DOI: 10.1159/000083340

Campaign to Control Genetic Blood Diseases in Bahrain

S. Al Arrayed

Salmaniya Medical Complex, Ministry of Health, Bahrain, Kingdom of Bahrain

Key Words

Bharain · Premarital counseling · Prevalence · Sickle cell anemia · β-Thalassemia

Abstract

Genetic blood diseases are frequent in Bahrain as in all Middle Eastern countries. Previous neonatal screening in 1984–1985 showed that the birth prevalence of sickle cell disease (SCD) was 2.1%, of sickle cell trait 11% and of glucose-6-phosphate dehydrogenase (G6PD) deficiency 25%. The Ministry of Health recognized the importance of controlling these diseases. In 1984, the first genetic clinic was established, which started the educational campaigns. Information booklets were prepared and distributed widely in schools and clubs in an attempt to increase awareness about these diseases among students and the public. In 1991, the Bahrain Hereditary Anemia Society was formed. In 1992, the Minister of Health formed a national committee for the prevention of genetic diseases in Bahrain. Screening of all pregnant women began, followed by newborn testing if the mother was found to be a carrier. In 1993, a premarital counseling (PMC) service was organized and in 1998, a student-screening project began. At this stage, we want to update the national birth prevalence figure by screening Bahraini newborns for these genetic diseases. This will help to design prevention programs and to measure the effect of health education on the previous birth prevalence figure. ***Subjects and Methods:*** A newborn screening study was conducted to determine the effects of this long-term campaign (16–18 years). Cord blood samples from 2,000 Bahraini newborns were tested for hemoglobinopathies and G6PD deficiency using HPLC. ***Results:*** 18 newborns were found to have SCD. The new birth prevalence figure for SCD in Bahrain is 0.9%, which indicates a 60% decline in the birth prevalence rate. ***Conclusion:*** With the continuation of education, awareness campaigns, screening of carriers and PMC, we expect the number of affected children born to be reduced tremendously over the next few years.

Introduction

The state of Bahrain is made up of an Archipelago of thirty-six islands, with an area of 694 km^2, located in the Arabian Gulf. The Kingdom of Saudi Arabia is on the west and Qatar on the east. The estimated 2002 population was 627,123. The percentage of Bahrainis is 62.4% and non-Bahrainis 37.6%.

Birth rate and population growth rates over the last 5 years have been stable. In 2002, the crude birth rate was 20.1 per 1,000 inhabitants. The infant mortality rate has been reduced to 7/1,000 live births in 2002. There has been a significant increase in life expectancy at birth. Children born today will on average live to 73.8 years [1].

KARGER

Fax +41 61 306 12 34
E-Mail karger@karger.ch
www.karger.com

1422–2795/05/0081–0052$22.00/0

Accessible online at:
www.karger.com/cmg

Dr. Shaikha Al Arrayed, MBChB, DHCG, PhD
Genetic Department, PO Box 12
Salmaniya Medical Complex, Ministry of Health
Bahrain (Kingdom of Bahrain)
Tel. +973 39607719, Fax +973 17689017, E-Mail ssarayed@batelco.com.bh

Falciparum malaria was endemic in Bahrain until 1970, and so the malaria-associated genetic defects of red cells such as sickle cell disease (SCD), thalassemia and glucose-6-phosphate dehydrogenase (G6PD) deficiency were expected to be common. These genetic diseases are passed from generation to generation through marriages between carriers. There is no real cure, and the management is supportive and symptomatic only. Preventive measures remain the best way of dealing with these diseases.

A newborn screening study conducted during 1984–1985 in Bahrain revealed that 2.1% had SCD (210 babies were born every year suffering from SCD and 11.2% were carriers for SCD) and 20% had G6PD deficiency in Bahrain [2]. The carrier status for β-thalassemia was found to be 2–4% in a premarital counseling (PMC) study [3].

These statistics reflect the fact that a large number of children are suffering from these diseases, thus affecting individuals, families and the society both socially and economically.

The Cyprus Thalassemia Control Program has succeeded in reducing the birth prevalence of β-thalassemia major in the country through measures such as health education, carrier screening, PMC and prenatal diagnosis. This success has encouraged other countries to adopt the same measures [4–8].

In Bahrain, during the 1970s patients attending the health services were treated symptomatically. Thalassemia patients were undertreated and undertransfused. Many of them developed complications and dysmorphic features. Many families had 3–5 children with thalassemia or SCD, and patients with SCD were treated as malingerers and subjected to neglect in emergency rooms and hospitals. They were often viewed as drug users since they often came for analgesics. At the beginning of the 1980s, the service for these patients improved. The Ministry of Health started to organize services provided through Health Centers, Medical Clinics and Pediatric Clinics. In 1984, the first Genetic Clinic was established at the Salmaniya Medical Complex. PMC started at the Genetic Clinic for families at risk. Information booklets were designed and distributed widely, and educational campaigns started in schools and clubs, to increase awareness about these diseases among students and the public.

In 1991 the Bahrain Hereditary Anemia Society was formed with the aim of fighting the hemoglobinopathies, and raising the standard of care for the patients. In 1992 the Minister of Health formed a National Committee for the Prevention of Genetic Diseases in Bahrain. The Committee formulated a plan aimed at preventing and eliminating these diseases. The Committee advised to concentrate on screening and education [9–12]. Screening all pregnant women started, and if the mother was found to be a carrier the newborn was also screened.

In 1993, PMC Services at Health Centers were organized by the Genetic Department and Maternal and Child Health Department. Training courses were given to all primary health care physicians, and a form was designed to standardize the service through all health centers. PMC became available on a wide basis through all the Health Centers. Initially acceptance was poor, but gradually it has gathered momentum [3].

In 1998, a student-screening project started. The aim of screening all students before they leave school was to raise awareness among the youth about hereditary anemia, and to advise them to take the right decision to prevent these diseases in their future families. A card with their blood test result was given to each student concomitant with an information booklet about the disease he carries.

This project is continuing for the 6th year, and it has had a tremendous effect on awareness, not only among students, but also among their families, as they ask to be tested if they know that their son is a carrier for a certain disease. It also positively affects attendance for PMC.

In 2001, the Shura Council agreed to make PMC mandatory before marriage. Genetic counseling will be provided for the couples at risk. These couples have the freedom of choice if they decide to marry against the advice. In 2002, the Ministers' Cabinet Council agreed on the PMC law, to be implemented within 1 year.

Subjects and Methods

At this stage we wished to (1) update the national figures for the birth prevalence of the common genetic blood disorders such as SCD, thalassemia, and G6PD among newborns in Bahrain, and compare it with the previously available figure; (2) establish a data base registry for all affected newborns; (3) identify the birth prevalence of other abnormal hemoglobin disorders; (4) study the parental ages among Bahraini parents, and (5) assess the consanguinity rate in Bahrain.

The study was a collaborative project between the Ministry of Health, the World Health Organization and the Bahrain Hereditary Anemia Society. The National Hereditary Disease Committee organized and directed the project, which was conducted between February 1 and April 30, 2002.

A total of 2,000 Bahraini newborns were tested, which represented all deliveries in the Ministry of Health Maternity Hospitals during the period of study. 80–90% of all Bahraini newborns during that period were included.

Table 1. Comparison of SCD frequency between newborns, mothers and fathers in 2002

	Mean age years	Total	SCD		Sickle cell carrier		Normal	
	n	n	n	%	n	%	n	%
Newborn	0	1,995	18	0.9	325	16.3	1,652	82.8
Mother	27	1,210	16	1.3	170	14.1	1,024	84.6
Father	42	260	5	1.9	21	8.1	234	90.0

Table 2. Comparison of the prevalence of SCD and sickle cell carrier frequency in 1986 and 2002

Year	Screened	Affected	Carriers
1986	10,327	217 (2.1%) (95% CI 1.8–2.4)	1,135 (11%) (95% CI 1.04–11.6)
2002	2,000	18 (95% CI 14.7–18.0)	326 (95% CI 0.5–1.42)
p		0.001	0.001

CI = Confidence interval.

Cord blood samples from the newborns were collected in tubes containing EDTA, and transferred to the Hematology Laboratory at the Salmaniya Medical Complex for analysis. High-performance liquid chromatography was used to perform electrophoresis on the cord blood samples, with a fluorescent screening technique for G6PD deficiency.

A questionnaire was completed for each newborn by nurses in the Maternity Units. The form included demographic data, parental age group, and consanguinity, and was coded for computerization. Senior personnel in each field undertook the requisite training of doctors, nurses and technicians. A confirmatory test was conducted on all the newborns with SCD.

The data were analyzed using the statistical package for social sciences [13]. The χ^2 test was used to compare frequencies between the carriers and affected patients in the years in question. $p < 0.05$ was considered as the cutoff value of significance. 95% confidence interval analysis was done for carriers and affected patients.

Results

Five samples from the newborns were excluded, because the blood was clotted. Only 0.9% (18/1,995) Bahraini newborns born in Ministry of Health facilities had SCD, while the sickle cell trait was found in 16.3% (325/1,995) of the newborns (table 1). There was a significant reduction in the birth prevalence of SCD between 1986 and 2002 (p = 0.001). Similarly, there was a significant increase in the frequency of carriers during that time (p = 0.001; table 2).

SCD was found in 1.9% (5/260) of the fathers and 1.3% (16/1,210) of the mothers tested. Sickle cell trait was found in 8% (21/260) of the fathers and 14% (170/1,210) of the mothers tested. G6PD deficiency was present in 18.0% of male and 10.0% of female newborns (table 3).

Table 3. The prevalence of G6PD deficiency in Bahraini newborns

Status	Male		Female		Total	
	n	%	n	%	n	%
Normal	644	81.2	661	88.9	1,305	85
Reduced	148	18.7	81	10.9	229	14.9
Trait	–	–	1	0.13	1	0.1

Newborn screening for G6PD (n = 1,535) 2002.

Discussion

The 0.9% SCD birth prevalence in 2002 compares with the previous estimate of 2.1% reported in 1984–1985, i.e. a decline of almost 60% ($p = 0.001$). Now we expect the birth of 90 babies annually with SCD, instead of 210, which means that 120 newborns have been protected from having the disease annually. The sickle cell trait prevalence of 16.3% in 2002 compares with previous estimates of 11% in 1986 ($p = 0.001$; table 2), 13% in student screening during 1998 and 14% in 2001. As shown in table 1, the incidence of SCD was higher in the parents than in their children. The trait rate showed the opposite trend, which had been expected as more carriers (HbA/HbS) marry (HbA/HbA) persons. With the continuation of education, awareness campaigns, carrier screening and PMC we expect the number of children born with SCD to be reduced tremendously in the next few years.

The prevalence of G-6-PD deficiency of 18% in the male newborns and 10% in the female newborns compares with reported values of almost 25% in many previous studies [2].

Although SCD is common in Bahrain, a previous study showed that we have the mild form of SCD in approximately 90% of patients. DNA analysis has proven the presence of the Asian haplotype as the common haplotype. This mutation is associated with a high level of HbF, a situation similar to that in the Eastern Province of Saudi Arabia [14, 15].

For the first time, Bahrain has recorded less than 1% of newborns with SCD. If this trend continues, Bahrain could soon become the first country in the Gulf region to completely eradicate the disease, and we hope to be able to protect all our children from this chronic and debilitating disease in the near future. This is a success story for both the Ministry of Health in Bahrain, the Bahrain Hereditary Anemia Society and all those involved in this campaign.

References

1 Health Statistics 2002: Bahrain Health Information Directorate, Ministry of Health, 2002.

2 Mohammed A, Al-Hilli F, Nadkarni K, Bhagwat G, Bapat J: Hemoglobinopathies and glucose-6-phosphate dehydrogenase deficiency in hospital birth in Bahrain. Ann Saudi Med 1992;12:536–539.

3 Al Arrayed S: Premarital counseling. East Mediterr Health J 1997;3:415–419.

4 Community Control of Hereditary Anaemia: Memorandum from a WHO meeting. Bull World Health Organ 1983;61:63–80.

5 Angastiniotis M, Kyriakidou S, Hadjiminas M: The Cyprus Thalassemia Control Program. Birth Defects 1988;23:417–432.

6 Angsstinoiotis M, Kyriakidou S, Hadjiminas M: How thalassemia was controlled in Cyprus. World Health Forum 1986;7:293–297.

7 Community Control of Genetic and Congenital Disorders: EMRO Technical Publication Series 24. Geneva, World Health Organization, 1997, pp 194–199.

8 Hereditary anaemia (genetic basis, clinical features, diagnosis and treatment). Bull World Health Organ 1982;60:643–660.

9 Al-Arrayed S: The nature of sickle-cell disease in Bahrain. J Bahrain Med Soc 1994;6:125–130.

10 Al Arrayed S, Hamza A: A survey of patients with sickle cell presentation to Accident and Emergency Department of SMC. Bahrain. J Bahrain Med Soc 1995;7:105–112.

11 Al-Arrayed SS, Haites N: Features of sickle-cell disease in Bahrain. East Mediterr Health J 1995;1:112–119.

12 Al-Arrayed S: Hematological characteristics in Bahraini sickle-cell disease patients. J Bahrain Med Soc 1991;2:32–35.

13 Norusis MJ: SPSS/PC+ for Windows, Base System and Advanced Statistic User's Guide, Windows version 11. Chicago, SPSS, 1998.

14 Al Arrayed SS: Beta globin gene haplotype in Bahraini sickle cell anemia patients. Bahrain Med Bull 1995;1:15–20.

15 Nadkarni KV, Al-Arrayed SS, Bapat JP: Incidence of genetic disorders of haemoglobin in the hospital population of Bahrain. Bahrain Med Bull 1991;131:19–24.

Community Genetics

Community Genet 2005;8:56–60
DOI: 10.1159/000083341

A Family Study in Oman: Large, Consanguineous, Polygamous Omani Arab Pedigrees

A Planned Model for the Study of Complex Diseases

M.O. Hassan[a] S. Albarwani[a] S. Al Yahyaee[b] S. Al Haddabi[a] S. Rizwi[a] A. Jaffer[b] J. Al-Lawati[b] G. Cai[c] Anthony G. Comuzzie[b] R.A. Bayoumi[a]

[a]College of Medicine, Sultan Qaboos University, and [b]Ministry of Health, Muscat, Sultanate of Oman; [c]Southwest Foundation, San Antonio, Tex., USA

Key Words
Complex diseases · Oman · Pedigrees

Abstract
Objective: To establish a suitable human model for the study of the genetics of complex diseases. ***Methods:*** We have selected an Omani Arab population to provide the statistical power required to study the genetics of complex diseases with confidence. This model consists of five multigenerational highly inbred pedigrees, descending from a small number of founders just a few generations ago with environmental homogeneity, restricted geographical distribution, detailed records and well-ascertained and -validated pedigrees. Stringent criteria were adopted for defining the phenotypes of hypertension, diabetes mellitus, dyslipidemias and obesity. The SOLAR genetic software package was used to draw the pedigree structure. ***Results:*** Outstanding statistical power to detect susceptibility loci was obtained. ***Conclusions:*** This model represents a large homogeneous human family-based population for the study of genetic and environmental factors contributing to complex diseases.

Introduction

Complex diseases result from the interactions of genetic and environmental factors. However, genetic studies are complicated by phenomena such as genetic heterogeneity, variable age at onset and clinical expression, incomplete penetrance, and wide-ranging, ill-understood, environmental triggers [1, 2]. Because genetic influences are less clear-cut, success in identifying genes implicated in the pathogenesis has been scanty, so far [3, 4]. The main difficulties associated with studies of complex traits is to achieve the necessary statistical power and reproducibility [5–7]. In heterogeneous populations, very large sample sizes are required to detect the effects of multiple genes, both for association and linkage studies of disease markers, and admixture in such populations leads to falsely high heritability estimates [8, 9].

The ideal population to provide the statistical power required to study complex diseases with confidence should therefore consist of multigenerational pedigrees, descended from a small number of founders just a few generations ago, and with environmental homogeneity, restricted geographical distribution, detailed records, well-ascertained and validated pedigrees, and inbreeding as a norm [10].

KARGER

Fax +41 61 306 12 34
E-Mail karger@karger.ch
www.karger.com

1422–2795/05/0081–0056$22.00/0

Accessible online at:
www.karger.com/cmg

Prof. M.O. Hassan
Chairman, Department of Physiology, College of Medicine and Health Sciences
Sultan Qaboos University, PO Box 35
Muscat 123 (Sultanate of Oman)
Tel. +968 515109, Fax +968 513419, E-Mail mhassan@squ.edu.om

We have chosen to study a homogeneous population of families with very large pedigree sizes (160–325 subjects) in some inland villages of the Nizwa region of Oman. The families converge towards common descendents of Arab origin, and the founders date back to no more than 200 years. Traditionally, each individual is genealogically well defined within the family, and records are well maintained by individuals and their elders. Often fathers, grandfathers and even great grandfathers are still alive. This population has maintained homogeneity by extremely high levels of inbreeding due to the tradition of encouraging consanguineous marriage, mostly between first cousins. Such unions occur in over 50% of marriages. Even when there is a lesser coefficient of inbreeding, the genetic homogeneity is also maintained by endogamy [11]. The relative isolation of these families in small villages is due to rough terrain. Travel, education and other modernizations of the area only developed in the past 20–30 years.

We believe this model has a better chance of achieving the power required to examine the genetic elements of complex diseases than many larger studies in outbred, heterogeneous populations. To add further precision, we have adopted very stringent criteria for defining the phenotypes, which are dissected down to all possible components. For example, we are studying blood pressure as a continuous variable, rather than segregating individuals as hypertensives and normotensives. The phenotypes that contribute to incidental blood pressure reading such as stroke volume, heart rate, peripheral resistance, baroreceptor sensitivity, sympathovagal balance, endothelial function and the relevant blood hormone levels were also studied in detail at rest and during laboratory stress tests.

Subjects and Methods

We chose to study a homogeneous Arab population made of five large, extended and highly consanguineous families. The numbers of subjects in these five pedigrees were 325, 160, 230, 281 and 279, totaling 1,275 subjects (table 1).

The Wilayat (State) of Nizwa, the home of these very large extended Arab Muslim families, is 150 km south of the capital Muscat. It is a mountainous region dotted by several oases and seasonal river beds. The Omani Arabs who live in the Wilayat of Nizwa are relatively isolated. They have lived around the oases of Nizwa for several hundreds of years with little contact with the peoples of the coast who, in contrast, had close associations with the Indian subcontinent and East Africa for over 800 years.

The 1970 renaissance of Oman, accompanying the oil boom, resulted in rapid population growth and a dramatic change in lifestyle. While family structure remained unchanged, there has been an expansion in family size over a relatively short period of time. Our study and studies in the same area [12] have shown relatively high prevalence for hypertension (21%), dyslipidemias (41%), type II diabetes mellitus (5%), glucose intolerance (12%), obesity (5%) and the metabolic syndrome (20%).

Coalition Building is an important process and was tailored carefully for each pedigree. It was crucial for the success of the project and was carried out cautiously to ensure the suitability of the tribe/family in question for the study and for their future coopera-

Table 1. Data collected from each subject

Anthropometric	Age, weight, height, waist and hip circumference, and percent body fat
Biochemical tests	Fasting blood glucose
Glucose	Blood glucose 2 h after 75 g oral glucose
Hormones	Blood insulin level (fasting and 2 h after 75 g glucose) Growth hormone, leptin, cortisol, thyroxine and thyroid-stimulating hormone
Chemistry	Serum albumin, total protein, alkaline phosphatase, uric acid, urea and creatinine Fasting blood lipids: total, low- and high-density lipoprotein cholesterol and triglycerides
Hemodynamics	Performed using non-invasive beat-to-beat blood pressure measurements and impedance cardiography incorporated in a state-of-the-art Task Force Monitor Unit (CNS systems) [13, 14]; the following tests were performed after 10 min of rest, after 3 min of a Stroop Color and cold pressor tests
	Beat-to-beat systolic, diastolic and mean blood pressure, heart rate, R-R intervals and their reactivity to stress [13] Stroke volume, cardiac output, total peripheral resistance, left ventricular ejection fraction and cardiac muscle contractility Baroreceptor sensitivity and efficiency [15] Sympathetic and parasympathetic function and sympathovagal balance were derived using power spectral analysis of R-R intervals and diastolic blood pressure [16]
Ambulatory blood pressure	24-hour ambulatory blood pressure monitoring (Schiller), blood pressure variability and percent drop during sleep
Endothelial function	Assessed non-invasively after iontophoresis of acetylcholine and sodium nitroprusside into the forearm skin and using a laser Doppler scanner to measure changes in skin blood flow (Moores instruments)
	Resting electrocardiogram

Table 2. Pedigree data

	Taymsa pedigree 1	Birkat Al-Moz pedigree 2	Radat Al-Busaid pedigree 3	Karsha pedigree 4 (US)	Farq pedigree 5 (US)
Generations	8	7	12	11	
Individuals	530	277	492	510	
Founders	70	71	99	–	
Nuclear families	140	88	177	–	
Mean inbreeding coefficient	0.009	0.005	0.010	–	–
Subjects sampled in the pedigree	325	160	230	281	279
Subjects sampled outside the pedigree	5	24	16	12	15

US = Under study. Taymsa, Birkat Al-Moz, Radat Al-Busaid, Karsha and Farq are villages in Oman.

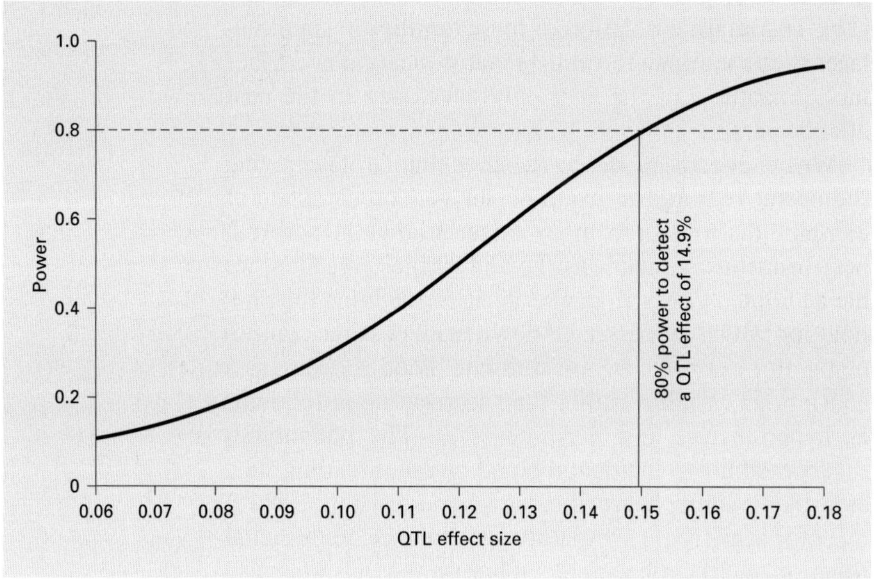

Fig. 1. Power analysis for linkage analysis (n = 1,275). QTL = Quantitative trait locus.

tion. The focus was mainly on the immediate and long-term benefits of the community from the project, especially from the clinical intervention trial, which caters for those with the above-indicated existing and emerging complex diseases as well as their follow up for at least 5–10 years. Results of genetic testing of benefit will be communicated to individuals observing strict confidentiality.

Village stations were selected within a walking distance or a short shuttle bus ride for participants in the study, while the main field research center, an annex of the Nizwa polyclinic, is situated within 20 km of all five villages. Both centers were used for 2-day detailed studies of the phenotypes and sample collection (table 2).

We performed an extensive series of simulations to evaluate the statistical power of localizing genes influencing quantitative factors associated with complex diseases in the five pedigrees. We designed our simulations so that the expected locus-specific effects were varied to account for 0–60% of the total phenotypic variation. One hundred replicates were performed for each set of generating values. Multipoint linkage analyses were performed on each replicate using SOLAR software (Southwest Foundation for Biomedical Research, San Antonio, Tex., USA). The results of these analyses were examined to determine what size of relative genetic effect could be reliably detected in a sample of 1,207 individuals in the five pedigrees. Classical statistical design of experiments generally requires that a power of 80% be targeted. Therefore, we determined the locus-specific heritability at which we will obtain 80% power to obtain a logarithm of odds (LOD) score greater than 3.

The infrastructure, capacity building, field work, data collection and the preliminary analysis of data were supported by strong local funding from His Majesty, the Sultan of Oman Research Trust Fund and the Ministry of Health. In recognition of the uniqueness of this study, the Mammalian Genotyping Services (Marshfield Clinic Research Foundation; funded by the National Heart, Lung and Blood Institute, USA) has agreed to genotype all 1,275 DNA samples using the Marshfield 800 marker 5 cM genetic map.

Results

All five pedigrees sampled claim different ancestry and carry different subtribal names, yet on closer examination they all seem to have common Arab ancestry, and intermarriage is common between them (table 1). Traditionally, each individual is genealogically well defined within a branched family, with well-maintained records of the individual, his father, grandfather and great grandfather, most of whom are still alive due the customary marriage at a very young age.

Figure 1 shows the statistical power curves as a function of quantitative-trait-locus-specific heritability. Using an LOD score of 3 as the critical cutoff point, we will have 80% power to locate a single specific locus that accounts for as little as 14.9% of the total phenotypic variation when we analyze the target sample of 1,275 individuals. It should be emphasized that this figure is very likely to be an underestimate of the true power, since we are currently working to link the five separate pedigrees into a single large pedigree, and have already been successful in making several such connections. However, one has to bear in mind that using an LOD score of 3 rather than 3.6 as the critical cutoff point is not ideal when dealing with complex traits as positive results would indicate suggestive linkage rather than significant linkage [17].

Discussion

A unique model is proposed for examining the genetics of hypertension, the dyslipidemias, obesity, diabetes and the metabolic syndrome. There are many features and strengths in this approach. We chose a homogeneous, genetically and geographically isolated population, consisting of pedigrees of very large size (160–325 subjects), with few recent founders (5th–10th generations), a very high degree of inbreeding, polygamy and excellent genealogy records. The tight founding population could probably result in a limited number of polymorphisms and mutations. In a series of simulations we evaluated the statistical power of this model to map genes influencing quantitative traits of complex diseases. The results have indicated that we have outstanding power to detect susceptibility loci in these pedigrees. The high prevalence of those complex diseases in this population will also provide added sufficient statistical power for linkage analysis. We have also adopted more stringent criteria for defining the phenotypes. Most importantly, all parameters measured were considered as continuous variables, and the phenotype itself is dissected down to all possible components, to come as close as possible to gene products. The average number of children for each marriage in this population is around six, and 60% of them are under 20 years of age. The population will also be useful for looking into precursors or predictors of complex diseases in children, the understanding of which may lead to new therapies to prevent their development or to mitigate their consequences later in life. Also, the numbers of children will probably double in size in the next 10 years, expanding each pedigree sufficiently for future studies.

Another major advantage is the guarantee of continuity of this study in the future. Recall of individuals and the repetition of experiments that need subject cooperation, e.g. to study hemodynamics, can easily be done. Agreement of the families to participate in a long-term study was almost unanimous. In conclusion, we believe that these Arab pedigrees will provide an important and powerful resource for the study of the genetics of complex diseases.

References

1 Rannala B: Finding genes influencing susceptibility to complex diseases in the post-genome era. Am J Pharmacogenomics 2001;1:203–221.

2 Lynch M, Conery JS: The origins of genome complexity. Science 2003;302:1401–1404.

3 Lesage S, Zouali H, Colombel JF, Belaiche J, Cezard JP, Tysk C, Almer S, Gassull M, Binder V, Chamaillard M, Le Gall I, Thomas G, Hugot JP: Genetic analyses of chromosome 12 loci in Crohn's disease. Gut 2000;47:787–791.

4 Tavtigian SV, Simard J, Teng DH, Abtin V, Baumgard M, Beck A, Camp NJ, Carillo AR, Chen Y, Dayananth P, et al: A candidate prostate cancer susceptibility gene at chromosome 17p. Nat Genet 2001;27:172–180.

5 Goldin LR, Chase GA: Improvement of the power to detect complex disease genes by regional inference procedures. Genet Epidemiol 1997;14:785–789.

6 Thomson G: Significance levels in genome scans. Adv Genet 2001;42:475–486.

7 Hirschhorn JN, Lohmueller K, Byrne E, Hirschhorn KA: A comprehensive review of genetic association studies. Genet Med 2002;4:45–61.

8 Yang H, Wang Y, Goldstein DR, Li Z, Vora H, Cantor RM: Factors influencing the identification of major genes in a complex disease genome scan. Genet Epidemiol 1997;14:933–938.

9 Rannala B: Gene genealogy in a population of variable size. Heredity 1997;78:417–423.

10 Arcos-Burgos M, Muenke M: Genetics of population isolates. Clin Genet 2002;61:233–247.
11 Bittles AH: Endogamy, consanguinity and community genetics (review). J Genet 2002; 81:91–8.
12 Al-Lawati JA, Mohammed AJ, Al Hinai HQ, Jousilahti P: Prevalence of the metabolic syndrome among Omani adults. Diabetes Care 2003;26:1781–1785.
13 Snieder H, Harshfield GA, Treiber FA: Heritability of blood pressure and hemodynamics in African and European-American youth. Hypertension 2003;41:1196–1201.
14 Parati G, Ongaro G, Bilo G, Glavina F, Castiglioni P, Di Rienzo M, Mancia G: Non-invasive beat-to-beat blood pressure monitoring: New developments (review). Blood Press Monit 2003;8:31–36.
15 Gollasch M, Tank J, Luft FC, Jordan J, Maass P, Krasko C, Sharma AM, Busjahn A, Bahring S: The BK channels B1 subunit gene is associates with human baroreflex and blood pressure regulation. J Hypertens 2002;20:927–933.
16 Beitzke M, Pfister P, Fortin J, Skrabal F: Autonomic dysfunction and hemodynamics in vitamin B12 deficiency. Auton Neurosci 2002; 97:45–54.
17 Lander E, Kruglyak L: Genetic dissection of complex traits: Guidelines for interpreting and reporting linkage results. Nat Genet 1995;11: 241–247.

Community Genetics

Community Genet 2005;8:61–64
DOI: 10.1159/000083342

Familial Predictors of Childhood Shyness: A Study of the United Arab Emirates Population

V. Eapen[a] R. Ghubash[a, b] M.O. Salem[a] S. Sabri[a]

[a]Department of Psychiatry, Faculty of Medicine and Health Sciences, United Arab Emirates University, Al Ain, United Arab Emirates; [b]Arabian Gulf University, Manama, Bahrain

Key Words
Arab culture · Childhood shyness · Social phobia

Abstract
Objectives: Phobic anxiety disorders generally breed true. In this regard, family studies have suggested an association between childhood shyness and maternal social phobia. In this study, the relationship between childhood shyness and maternal social anxiety was examined. ***Subjects and Methods:*** 203 5-year-old children from an Arabian Gulf community and their mothers were evaluated. The children were assessed in stage 1 using the shyness scale of Stevenson-Hinde and Glover and the Preschool Behavior Checklist (PBCL), while mothers completed the Liebowitz Social Anxiety Scale. In the second stage, blind clinical interviews were carried out to ascertain the diagnosis of psychiatric diagnoses, if any, using DSM-IV criteria. ***Results:*** 27% of the children were identified as shy using the shyness scale of Stevenson-Hinde and Glover and 19% scored above the cutoff for behavioral disturbance on the PBCL. Child shyness was associated with female gender and maternal social anxiety as indicated by scores on the Liebowitz Social Anxiety Scale, while an inverse relationship was noted with behavioral disturbance. The odds of the child having a high shyness score was increased if the mother had social anxiety (odds ratio = 2.14) and the child lived in a family that was 'not socially active' (odds ratio = 1.42). ***Conclusion:*** Our initial findings suggest that there may be a complex interaction between maternal social anxiety and family sociability in childhood shyness. Prospective longitudinal work is indicated.

Introduction

The lifetime prevalence of social phobia in adults is estimated to be 7–13% in western countries [1], while that of the social anxiety disorder in children and adolescents ranges from 0.5 to 4% [2]. Social phobia is a common, chronic disorder with significant negative consequences, including an increased rate of suicide attempts, antisocial behavior, impaired school performance during adolescence, impaired medical health, increased help-seeking behavior, poor employment performance, reduced social interaction and impaired social support [3]. Given such

KARGER

Fax +41 61 306 12 34
E-Mail karger@karger.ch
www.karger.com

1422–2795/05/0081–0061$22.00/0

Accessible online at:
www.karger.com/cmg

Dr. Valsamma Eapen, MBBS, DPM, PhD, MRCPsych
Faculty of Medicine and Health Sciences, UAE University
PO Box 17666
Al Ain (United Arab Emirates)
Tel. +971 3 7672000, Fax +971 3 7672995, E-Mail veapen@uaeu.ac.ae

varied effects of this disorder on the functioning and quality of life, identifying the risk factors and developing preventive strategies should be given appropriate priority. In this regard, identifying childhood characteristics and risk factors for the development of social phobia has important clinical implications.

There is significant research evidence to suggest that social phobia runs in families and that this familial tendency is genetic in origin [4]. It has also been shown that behavioral inhibition in childhood increases the risk of social phobia in adolescence [5, 6]. Furthermore, Cooper and Eke [7] suggested a specific association between childhood shyness and maternal social phobia. However, since sociability and behavioral characteristics are influenced by cultural norms, child-rearing practices and social ideals, we felt it would be interesting to explore the association between childhood shyness and maternal social phobia in an Arabian Gulf population.

Subjects and Methods

A representative random sample of 203 5-year-old children attending the Kindergarten facility of three government schools in the United Arab Emirates were studied. The mothers completed the shyness scale of Stevenson-Hinde and Glover [8] about their children, and a member of the school staff who knew the child well completed the Preschool Behavior Check List (PBCL) [9] for each child. The mothers also completed the Liebowitz Social Anxiety Scale about themselves. A subsample of children who scored above the cutoff score of 4 or more on the shyness scale and a randomly selected equal number of those who scored below 4 were interviewed by a psychiatrist, who was blind to the child's shyness score. Both the child and the mother were evaluated using DSM-IV (American Psychiatric Association) criteria to determine whether they met the criteria for the social anxiety disorder of childhood and adult social phobia, respectively.

Results

A total of 203 children were studied, comprising 116 girls and 87 boys. Of the 203 children, 55 (27.1%) were identified as shy, with a score of 4 or more on the shyness questionnaire. No major discrepancies were noted between the mother's report on the shyness scale and the school professional's opinion regarding the two items on the PBCL concerned with social inhibition. Of those who were identified as shy, there were 37 girls and 18 boys, with a female to male ratio of 2:1. On the PBCL, 38 (18.7%) children scored above the cutoff indicating significant behavioral disturbance. There were 23 boys and 15 girls in this group, with a male to female ratio of 1.5:1.

Table 1. Correlates of childhood shyness

Variable	Shy (n = 55)	Normal (n = 148)	p
Gender			
Male	18	70	0.048
Female	37	78	
Maternal social anxiety			
LSAS mean score for social fears	41.8	35.8	0.001
LSAS mean score for avoidance	41.6	36.0	0.012

LSAS = Liebowitz Social Anxiety Scale.

Using the Kruskal-Wallis test, child shyness was found to be significantly associated with social anxiety in the mother with respect to the Liebowitz Social Anxiety Scale for both the social fear (p = 0.001) and the avoidance (p = 0.012) components (table 1). There was an inverse relationship between behavioral disturbance, as indicated by PBCL scores and childhood shyness (0.003) as well as maternal social anxiety (0.035).

As indicated in table 1, there was a significant association between gender and shyness, with a female preponderance (p = 0.048). For behavioral disturbance on PBCL, a reverse association was noted for gender (p = 0.012), with a male preponderance.

Of the 55 children who scored 4 or above, every alternate child in the list was invited for the clinical interview. Of the 27 children thus selected, 20 were successfully interviewed, a response rate of 74.0%. 20 children from the low-score group were also interviewed in the second phase of the study. Of the 40 children thus interviewed, 3 mother-child pairs with a diagnosis of social phobia and social anxiety disorder, respectively, were identified. Two further children also scored positive for a diagnosis of social anxiety; the mother of one had subclinical social phobia and the second mother received no diagnosis. Of the 5 children who received a diagnosis of social anxiety disorder, all reported negatively to the question about family sociability (for example, entertaining guests at home). The odds of the child having a high shyness score was increased, if the mother had social anxiety (odds ratio = 2.14), and if the family was reported to be 'socially not active' (odds ratio = 1.42).

Discussion

Our findings suggest that there is an association between childhood shyness and maternal social phobia. A number of studies have suggested a familial mechanism underlying childhood shyness and adult social phobia, and have postulated a genetic contribution [10, 11]. Lieb et al. [12] found a strong association between parental social phobia and social phobia among their offspring. However, familial environmental factors have also been observed to be important, with a moderately high correlation for shyness between adoptive parents and their adopted children [13]. Furthermore, perceived parenting style has been found to be associated with the child's social phobia [12]. Thus child-rearing practices and shared family environment between the shy child and the mother with social phobia may be mediating factors in the genesis and maintenance of these conditions. Also, in our study, shared family environmental factors were noted as important. In this regard, it is interesting to note our finding of an increased risk of shyness in those children whose family, and in particular the mothers, were not socially active. This is in keeping with the observation that family sociability, as reported by the children and the mothers, is an important correlate for the development of social anxiety disorders in children [14]. Thus it would seem that living in a socially active family environment might act as a mitigating factor in the development of the social anxiety disorder in childhood.

We found the rates of shyness to be high among girls, and it has been suggested that some childhood risk factors may interact with gender to influence the development of social phobia [15]. On the contrary, we found higher rates among boys for behavioral disturbance and a negative association between shyness and behavioral disturbance. This is consistent with the findings of Biederman et al. [6], who noted that non-inhibited children were significantly more likely to have disruptive behavior disorders than inhibited or shy children.

Heiser et al. [16] reported that, while the majority of shy children are not socially phobic, there was a correlation between the severity of shyness and the presence of social phobia. Thus, a positive family history and severity of childhood shyness seem to be important predictors of the future course of this condition. It has also been found that the children of social phobic parents have increased rates of anxiety disorders in general [17]. While it is generally believed that nongenetic factors have only a limited effect on the development of social phobia, familial factors such as parental modelling of childhood social fears may be important in the expression of social phobia in children who are genetically susceptible [18]. In this regard, Kendler et al. [19] found that social phobia results from the combined effects of a slightly stronger genetic influence and non-specific environmental experiences.

While prediction of childhood shyness to adolescent anxiety disorder is modest, with most shy children in the community not developing an anxiety disorder [20], this is not the case for those with a positive family history. Thus it seems that for those with a genetic vulnerability, familial environmental factors may have an additive influence. This may have important implications for selecting children at risk for preventive intervention.

Although we found an association between childhood shyness and maternal social phobia, conclusions about causality cannot be convincingly drawn because of the cross-sectional nature of the study. Yet another limitation of this study is that we used the reports of the mothers in the initial phase of the study, leading to the possibility of a reporting bias. However, the fact that our observations remained consistent in the second phase of the study using a 'blind' clinical interviewer suggests that the findings are robust. Future prospective longitudinal studies are indicated, to elucidate the role of genetic versus shared environmental risk factors in the genesis of social anxiety disorders in childhood.

Acknowledgments

The authors would like to thank Dr. Abdel Azim Mabrouk, Pediatrician, and Dr. Salem Bin-Othman, Director of the School of Health Services, Al Ain, and Ms. Jawaher Al Khatebi, Ministry of Education, Dubai, for their administrative assistance in carrying out this study.

References

1 Furmark T: Social phobia: Overview of community surveys. Acta Psychiatr Scand 2002; 105:84–93.

2 Chavira D, Stein MB: Recent developments in child and adolescent social phobia. Curr Psychiatry Rep 2000;2:347–352.

3 Davidson JR, Hughes DL, George LK, Blazer DG: The epidemiology of social phobia: Findings from the Duke Epidemiological Catchment Area Study. Psychol Med 1993;23:709–718.

4 Fyer AJ, Mannuzza S, Chapman TF, et al: Specificity in familial aggregation of phobic disorders. Arch Gen Psychiatry 1995;52:564–573.

5 Hayward C, Killen JD, Kraemer HC, Taylor CB: Linking self-reported childhood behavioral inhibition to adolescent social phobia. J Am Acad Child Adolesc Psychiatry 1998;37:1308–1316.

6 Biederman J, Hirshfeld-Becker DR, Rosenbaum JF, Herot C, Friedman D, Snidman N, Kagan J, Faraone SV: Further evidence of association between behavioral inhibition and social anxiety in children. Am J Psychiatry 2001;158:1673–1679.

7 Cooper PJ, Eke M: Childhood shyness and maternal social phobia: A community study. Br J Psychiatry 1999;174:439–443.

8 Stevenson-Hinde J, Glover A: Shy girls and boys: A new look. J Child Psychol Psychiatry 1996;37:181–187.

9 McGuire J, Richman N: The genetic epidemiology of phobias in women. Arch Gen Psychiatry 1986;49:273–281.

10 Kendler KS, Neale MC, Kessler RC, et al: The genetic epidemiology of phobias in women. Arch Gen Psychiatry 1992:49:273–281.

11 Fyer AJ: Heritability of social anxiety: A brief review. J Clin Psychiatry 1993;54(suppl):10–12.

12 Plomin R, DeFries JC: Origins of Individual Differences in Infancy: The Colorado Adoption Project. New York, Academic Press, 1985.

13 Lieb R, Wittchen HU, Hofler M, Fuetsch M, Stein MB, Merikangas KR: Parental psychopathology, parenting styles, and the risk of social phobia in offspring: A prospective-longitudinal community study. Arch Gen Psychiatry 2000;57:859–866.

14 Bogels S, van Oosten A, Murais P, Smulders D: Familial correlates of social anxiety in children and adolescents. Behav Res Ther 2001; 39:273–287.

15 Chartier MJ, Walker JR, Stein MB: Social phobia and potential childhood risk factors in a community sample. Psychol Med 2001;31: 307–315.

16 Heiser NA, Turner SM, Beidel DC: Shyness: Relationship to social phobia and other psychiatric disorders. Behav Res Ther 2003;41:209–221.

17 Mancini C, van Ameringen M, Szatmari P, Fugere C, Boyle M: A high-risk pilot study of the children of adults with social phobia. J Am Acad Child Adolesc Psychiatry 1996;35:1511–1517.

18 Stein MB, Chavira DA, Jang KL: Bringing up bashful baby. Developmental pathways to social phobia. Psychiatr Clin North Am 2001;24: 661–675.

19 Kendler KS, Neale MC, Kessler RC, Heath AC, Eaves LJ: The genetic epidemiology of phobias in women. The interrelationship of agoraphobia, social phobia, situational phobia and simple phobia. Arch Gen Psychiatry 1992; 49:273–281.

20 Prior M, Smart D, Sanson A, Oberklaid F: Does shy-inhibited temperament in childhood lead to anxiety problems in adolescence? J Am Acad Child Adolesc Psychiatry 2000;39:461–468.

Community Genetics

Author Index Vol. 8, No. 1, 2005

Fax + 41 61 306 12 34
E-Mail karger@karger.ch
www.karger.com

Accessible online at:
www.karger.com/cmg

Community Genetics

Subject Index Vol. 8, No. 1, 2005

KARGER
Fax + 41 61 306 12 34
E-Mail karger@karger.ch
www.karger.com

Accessible online at:
www.karger.com/cmg